AF410702

Cheese and Whey

Cheese and Whey

Editors

Golfo Moatsou
Ekaterini Moschopoulou

MDPI • Basel • Beijing • Wuhan • Barcelona • Belgrade • Manchester • Tokyo • Cluj • Tianjin

Editors
Golfo Moatsou
Agricultural University of
Athens
Greece

Ekaterini Moschopoulou
Agricultural University of
Athens
Greece

Editorial Office
MDPI
St. Alban-Anlage 66
4052 Basel, Switzerland

This is a reprint of articles from the Special Issue published online in the open access journal *Foods* (ISSN 2304-8158) (available at: https://www.mdpi.com/journal/foods/special_issues/Cheese_Whey).

For citation purposes, cite each article independently as indicated on the article page online and as indicated below:

LastName, A.A.; LastName, B.B.; LastName, C.C. Article Title. *Journal Name* **Year**, *Volume Number*, Page Range.

ISBN 978-3-0365-4955-2 (Hbk)
ISBN 978-3-0365-4956-9 (PDF)

© 2022 by the authors. Articles in this book are Open Access and distributed under the Creative Commons Attribution (CC BY) license, which allows users to download, copy and build upon published articles, as long as the author and publisher are properly credited, which ensures maximum dissemination and a wider impact of our publications.
The book as a whole is distributed by MDPI under the terms and conditions of the Creative Commons license CC BY-NC-ND.

Contents

About the Editors

Golfo Moatsou

Golfo Moatsou is a Professor in the Department of Food Science and Human Nutrition in the Agricultural University of Athens, Greece. Her research and teaching activities focus on dairy and cheese science and technology. She is a single or co-author of 130 publications, book chapters, and books. Moreover, she has delivered several lectures in workshops, seminars, and short courses and has been Guest Editor of 4 special issues of SCI journals. She is a member of the Editorial Board of 4 and reviewer of more than 20 SCI journals. Most of her research activities have been carried out in the frame of research projects funded by state agencies and dairy companies.

Ekaterini Moschopoulou

Ekaterini Moschopoulou (Assistant Professor) completed her BSc degree at the Agricultural University of Athens (AUA), MSc degree in Dairy Science at Reading University (UK) and PhD degree at AUA. She teaches theoretical and lab courses of dairy technology to graduate and postgraduate students. Her research interests include the milk-clotting enzymes, cheese, yoghurt, and ice cream technology, membrane technology, implementation of modern processing methods in milk and dairy products, development of new dairy products, milk heat treatment, and valorization of whey/acid whey. She is a referee and guest editor in various scientific journals in the field of dairy science and technology.

Editorial

CHEESE and WHEY: The Outcome of Milk Curdling

Golfo Moatsou * and Ekaterini Moschopoulou

Laboratory of Dairy Research, Department of Food Science and Human Nutrition, Iera Odos 75, 11855 Athens, Greece; catmos@aua.gr
* Correspondence: mg@aua.gr; Tel.: +30-210-529-4630

Citation: Moatsou, G.; Moschopoulou, E. CHEESE and WHEY: The Outcome of Milk Curdling. *Foods* **2021**, *10*, 1008. https://doi.org/10.3390/foods10051008

Received: 29 April 2021
Accepted: 1 May 2021
Published: 5 May 2021

Publisher's Note: MDPI stays neutral with regard to jurisdictional claims in published maps and institutional affiliations.

Copyright: © 2021 by the authors. Licensee MDPI, Basel, Switzerland. This article is an open access article distributed under the terms and conditions of the Creative Commons Attribution (CC BY) license (https://creativecommons.org/licenses/by/4.0/).

The present Special Issue is dedicated to both products of the cheesemaking process, that is cheese and whey. Cheese is an excellent and complex food matrix that preserves in concentrated form valuable milk constituents, such as proteins, minerals, vitamins, and biofunctional lipids. The formation of cheese mass requires the removal of whey, i.e., water and soluble milk substances—proteins, minerals, lactose, and vitamins—. According to Fox and McSweeney [1], cheese is the most diverse group of dairy products manufactured from a few kinds of milk by means of a protocol that is more or less common in respect to the first 24 h of manufacture. They suggest that cheese is the most interesting and challenging dairy product due to an inherent instability that results from a series of biochemical changes. Whey is the valuable byproduct derived from the cheesemaking process. Lactose and whey proteins are the main compounds of whey and the main reasons for its valorization, through the production of whey cheeses, functional/nutritional whey proteins concentrate, bioactive peptides, and oligosaccharides [2,3].

In the present article collection, two publications for the Parmigiano Reggiano cheese are included. Considering the demanding cheese making conditions and the long-term ripening of this cheese variety, relevant studies are of particular importance. Franceschi et al. [4] designed a two-year study in various cheese factories to assess the performance of milk with high somatic cell count (SCC), i.e., from 400,000 to 1,000,000, when it is used for the manufacture of Parmigiano Reggiano cheese. Compared to control milk with SCC < 400,000 cells/mL, the high SCC milk contained significantly lower casein, phosphorous, and chloride and exhibited lower casein number and titratable acidity. Statistically significant reduction of yield by approximately 8.8% was observed after 24 months of ripening of cheese made from milk with high SCC although cheese moisture was not affected. In fact, the reduction was significant for both actual and dry yield, being in accordance with the significantly higher fat loss—by approximately 20%—which occurred during cheesemaking of high SCC milk. According to the findings, the inferior composition of high SCC milk and the curd behavior made there from, jeopardize among other things, the financial result of this laborious production process.

D'Incecco et al. [5] studied the effect of extending the ripening period from 12 to 50 months on the Parmigiano Reggiano cheese characteristics. It was shown that in this extra-hard cheese, moisture content decreased constantly throughout the ripening period. Primary proteolysis was due to chymosin and mostly to plasmin activity. Secondary proteolysis revealed high numbers of FAA with glutamic acid being the predominant. Regarding volatile compounds, short chain odd-numbered FFA, esters, and ketones were the most abundant without, however, causing any off-flavor. Cheese microstructure showed that free fat was trapped in the protein network. Moreover, the cheese porosity increased significantly, and tyrosine crystals appeared after the 12 months of ripening. The researchers concluded that both biochemical and structural changes continue to take place in this cheese up to 50 months of ripening and, they suggest that if this fact is communicated to consumers, the 'long-ripened' cheese could achieve a better price.

A particular hard-type Italian PDO cheese named Nostrano Valtrompia manufactured from raw milk partially skimmed by spontaneous creaming and supplemented with saffron is the objective of the article written by Bettera et al. [6]. The article presents the key physicochemical, macrostructural, and sensory properties of this cheese type and compares the effect of a 12 to 16 month ripening period in a non conditioned traditional cellar with ripening in a temperature conditioned warehouse. Interestingly, separate samples were taken from the rind, the underrind and the inner part of cheeses for the estimation of moisture, aw, and color parameters. The moisture gradient from the rind to the center of the cheese, —increasing from 15.3 to 31%, on average—influenced texture, moisture, aw, and color. Although cheeses ripened under the two different conditions shared a common identity, cheeses ripened in a conditioned warehouse exhibited a slightly softer texture, a slightly different porosity distribution, and a different sensory perception compared to traditionally ripened counterparts. Based on their findings, the authors propose the use of new sites in the region that can be used as alternative ripening warehouses, to overcome the "capacity" difficulties in the PDO area.

The objective of the study of Štefániková et al. [7] was the raw or pasteurized soft/creamy sheep milk traditional Bryndza cheese of Slovakia, which has a limited period of ripening, a moisture ranging from 52 to 56%, and a pH from 5.1 to 5.3. They examined the effect of the manufacturing way—traditional or industrial—on cheese gross composition, microflora, and volatile organic compounds; the latter by means of e-nose and GC-MS. Season and production conditions affected the features of Bryndza cheese. Presumptive lactococci, followed by presumptive lactobacilli and enterococci were the most dominant bacterial groups, whereas Dipodascus geotrichum was always detected. As expected, the lack of heat treatment of milk dramatically increased coliforms and enterococci counts. Key volatile organic compounds were ethyl acetate, isoamyl acetate, 2-butanone, hexanoic acid, D-limonene, and 2,3-butanedione, significantly affected by the production conditions and season. For example, according to PCA analysis of the e-nose dataset, cheeses produced in May are differentiated from those produced in September.

The results of the analysis of water-soluble extracts of Gouda-type cheeses, made with different starters by nano ultra-high-performance chromatography data-independent acquisition high-resolution mass spectrometry (nanoUHPLC-DIA-HRMS), are presented in the article of Arju et al. [8]. They proposed a cheese peptide method with a gradient time shorter than usual, by which they identified 558 cheese peptides throughout 90 days of ripening. They concluded that the steps of sample preparation should be further investigated to enhance transmission of shorter peptides related to bioactivity and flavor.

In contrast to ripened cheeses, fresh spreadable cheese is a special cheese category usually derived from the combination of renneting and acid coagulation of the cheese milk. Several technological parameters may affect the characteristics of such a cheese product. Fat-content, homogenization, and heat treatment of the cheese milk for producing Quark-type cheese were studied by Lepesioti et al. [9]. The researchers reported that homogenization increased the moisture and protein retention in cheese, and this was more pronounced in full-fat cheese. Moreover, cheeses made with homogenized milk were softer and less adhesive, gummy, and chewy from those made with unhomogenized milk. Heat treatment also improved both, the moisture and protein retention in cheese. The authors finally concluded that heat treatment under conditions that denature β-Lg and α-La by 8% and 25%, respectively, is applied to the cheese milk, is consistent with the development of an appropriate texture for this cheese type without adding stabilizers; hence, a 'clean label' full- or reduced-fat Quark-type cheese can be produced.

In the framework of the circular economy, whey can be returned to cheese milk to improve the characteristics of reduced fat cheese. This is due to the ability of whey proteins particles to act as fat replacers. Borges et al. [10] used heat treated (90 °C for 20 min) UF concentrated whey and other dairy byproducts, i.e., buttermilk and second cheese whey (Sorelho), in the manufacture of reduced fat wash curd bovine cheese. They found that the fortification of the cheese milk with concentrated whey at a ratio of 5% increased the protein content as well as the ratio of protein in dry matter and fat in dry matter of cheese, which was ripened up to 90 days. In parallel, it increased the pH, the hardness, and chewiness and decreased the adhesiveness of cheese after 30 days of ripening. The researchers reported that further investigation is needed regarding the form of the added byproducts, i.e., liquid or dry form as well the optimization of mixtures of them.

Hussein et al. [11] investigated the hydrolysis of WPC by alcalase by means of response surface methodology. The aim of the study was to determine the appropriate conditions for the production of protein hydrolysates with dual biofunctionalities, that is, angiotensin-I converting enzyme (ACE) inhibitory and antioxidant activities. The optimized hydrolysis parameters were: temperature 58.2 °C, E/S ratio 2.5%, pH 7.5, and duration 361.8 min. Under these conditions, the experimental results were similar to the respective predicted values, i.e., 89.2% hydrolysis degree vs. theoretical 89.6%, 98.4% ACE-inhibitory activity vs. 98.8%, 50.1% DPPH radical scavenging activity vs. 50.6%, and 73.1% ferrous ion chelating activity vs. 74.0%. The authors suggested that their findings could be further exploited in research studies for the in vivo efficacy of WPC hydrolysate.

Whey is used in beverages to increase their nutritional and biofunctional value. Whey protein isolate (WPI) has high nutritional value because it contains more than 90% whey protein and for this reason it is used in products aiming to special diets. However, WPI in whey beverages with pH 4–6.0 exhibits instability since the pI of whey proteins is around 5.1. Shang et al. [12] achieved to improve the stability of WPI in an aqueous solution at pH 4.5 by conjunction with polysaccharide from Flammulina velutipes, a species of edible mushroom, via noncovalent interactions. This polysaccharide (FVPs) interacting with WPI alters the secondary structure of whey proteins, and through electrostatic interactions with the protein surface, improves the protein stability. The ratio 1:12.5 between FVPs and WPI was reported as the optimum for noncovalent interactions. The complex WPI-FVPs showed increased antioxidant activity regarding the ABTS-, hydroxyl- and superoxide anion-radical inhibition. Moreover, it remained intact during in vitro digestibility by pepsin, while it was completely hydrolyzed by trypsin. The researchers concluded that the WPI-FVPs complex could be used in WPI-based beverages made by the food industry.

León-López et al. [13] investigated the characteristics of whey-based fermented beverages fortified with hydrolyzed collagen at ratios from 0.3% to 1%. It was shown that addition of collagen increased the pH, the viscosity, the nutritional value, and the in vitro bioavailability of the product. Specifically, addition of 1% collagen increased protein content and the antioxidant activity regarding the ABTS and DPPH radical inhibition. Moreover, no pathogen microorganism was detected. The authors proposed that this product could be addressed to specific consumers groups such as elderly people and athletes.

Concluding the presentation of this Special Issue, we would like to thank the above-mentioned research teams for their contributions in the present article collection, which provide excellent examples for the multidisciplinary character of the research on cheese and whey.

Funding: This research received no external funding.

Conflicts of Interest: The authors declare no conflict of interest.

References

1. Fox, P.F.; McSweeney, P.L.H. Cheese an overview. In *Cheese: Chemistry, Physics and Microbiology*, 4th ed.; McSweeney, P.L.H., Fox, P.F., Cotter, P.D., Everett, D.W., Eds.; Academic Press: London, UK, 2017; pp. 5–22.
2. Yadav, J.S.S.; Yana, S.; Pilli, S.; Kumar, L.; Tyagi, D.D.; Surampalli, R.Y. Cheese whey: A potential resource to transform into bioprotein, functional/nutritional proteins and bioactive peptides. *Biotechnol. Adv.* **2015**, *33*, 756–777. [CrossRef] [PubMed]
3. Barba, F.J. An Integrated Approach for the Valorization of Cheese Whey. *Foods* **2021**, *10*, 564. [CrossRef] [PubMed]
4. Franceschi, P.; Faccia, M.; Malacarne, M.; Formaggioni, P.; Summer, A. Quantification of Cheese Yield Reduction in Manufacturing Parmigiano Reggiano from Milk with Non-Compliant Somatic Cells Count. *Foods* **2020**, *9*, 212. [CrossRef] [PubMed]
5. D'Incecco, P.; Limbo, S.; Hogenboom, J.; Rosi, V.; Gobbi, S.; Pellegrino, L. Impact of Extending Hard-Cheese Ripening: A Multiparameter Characterization of Parmigiano Reggiano Cheese Ripened up to 50 Months. *Foods* **2020**, *9*, 268. [CrossRef] [PubMed]
6. Bettera, L.; Alinovi, M.; Mondinelli, R.; Mucchetti, G. Ripening of Nostrano Valtrompia PDO Cheese in Different Storage Conditions: Influence on Chemical, Physical and Sensory Properties. *Foods* **2020**, *9*, 1101. [CrossRef] [PubMed]

7. Štefániková, J.; Ducková, V.; Miškeje, M.; Kačániová, M.; Čanigová, M. The Impact of Different Factors on the Quality and Volatile Organic Compounds Profile in "Bryndza" Cheese. *Foods* **2020**, *9*, 1195. [CrossRef] [PubMed]
8. Arju, G.; Taivosalo, A.; Pismennoi, D.; Lints, T.; Vilu, R.; Daneberga, Z.; Vorslova, S.; Renkonen, R.; Joenvaara, S. Application of the UHPLC-DIA-HRMS Method for Determination of Cheese Peptides. *Foods* **2020**, *9*, 979. [CrossRef] [PubMed]
9. Lepesioti, S.; Zoidou, E.; Lioliou, D.; Moschopoulou, E.; Moatsou, G. Quark-Type Cheese: Effect of Fat Content, Homogenization, and Heat Treatment of Cheese Milk. *Foods* **2021**, *10*, 184. [CrossRef] [PubMed]
10. Borges, A.R.; Pires, A.F.; Marnotes, N.G.; Gomes, D.G.; Henriques, M.F.; Pereira, C.D. Dairy by-Products Concentrated by Ultrafiltration Used as Ingredients in the Production of Reduced Fat Washed Curd Cheese. *Foods* **2020**, *9*, 1020. [CrossRef] [PubMed]
11. Hussein, F.A.; Chay, S.Y.; Zarei, M.; Auwal, S.M.; Hamid, A.A.; Wan Ibadullah, W.Z.; Saari, N. Whey Protein Concentrate as a Novel Source of Bifunctional Peptides with Angiotensin-I Converting Enzyme Inhibitory and Antioxidant Properties: RSM Study. *Foods* **2020**, *9*, 64. [CrossRef] [PubMed]
12. Shang, J.; Liao, M.; Jin, R.; Teng, X.; Li, H.; Xu, Y.; Zhang, L.; Liu, N. Molecular Properties of Flammulina velutipes Polysaccharide–Whey Protein Isolate (WPI) Complexes via Noncovalent Interactions. *Foods* **2021**, *10*, 1. [CrossRef] [PubMed]
13. León-López, A.; Pérez-Marroquín, X.A.; Campos-Lozada, G.; Campos-Montiel, R.G.; Aguirre-Álvarez, G. Characterization of Whey-Based Fermented Beverages Supplemented with Hydrolyzed Collagen: Antioxidant Activity and Bioavailability. *Foods* **2020**, *9*, 1106. [CrossRef] [PubMed]

Article

Quark-Type Cheese: Effect of Fat Content, Homogenization, and Heat Treatment of Cheese Milk

Sofia Lepesioti, Evangelia Zoidou, Dionysia Lioliou, Ekaterini Moschopoulou and Golfo Moatsou *

Laboratory of Dairy Research, Department of Food Science and Human Nutrition, Agricultural University of Athens, Iera Odos 75, 118 55 Athens, Greece; sofialepesioti@gmail.com (S.L.); ezoidou@aua.gr (E.Z.); dionisia.lioliou@gmail.com (D.L.); catmos@aua.gr (E.M.)
* Correspondence: mg@aua.gr; Tel.: +30-210-529-4630

Abstract: The effect of homogenization and fat reduction in combination with variable heating conditions of cow milk on the characteristics of Quark-type cheese were investigated. The mean composition of full-fat cheeses was 71.96% moisture, 13.95% fat, and 10.31% protein, and that of its reduced-fat counterparts was 73.08%, 10.39%, and 12.84%, respectively. The increase of heat treatment intensity increased moisture retention and improved the mean cheese protein-to-fat ratio from 0.92 to 1. Homogenization increased the moisture and protein retention in cheese, but the effect was less intense for milk treated at 90 °C for 5 min. The extended denaturation of whey proteins resulted in harder, springier, and less cohesive cheese ($p < 0.05$). Treatment of milk at 90 °C for 5 min resulted in higher residual lactose and citric acid and lower water-soluble nitrogen contents of cheese ($p < 0.05$); the latter was also true for homogenization ($p < 0.05$). Storage did not affect the composition and texture but decreased galactose and increased citric acid and soluble nitrogen fractions ($p < 0.05$). In conclusion, heat treatment conditions of milk that induced a considerable denaturation of β-lactoglobulin and left a considerable amount of native α-lactalbumin was adequate for the manufacture of a "clean-label" Quark-type cheese, whereas homogenization was more effective for full-fat cheese.

Keywords: Quark-type cheese; cow cheese milk homogenization; cheese milk heat treatment; reduced-fat cheese; sugars and organic acids; proteolysis indices; texture profile analysis; whey protein denaturation

Citation: Lepesioti, S.; Zoidou, E.; Lioliou, D.; Moschopoulou, E.; Moatsou, G. Quark-Type Cheese: Effect of Fat Content, Homogenization, and Heat Treatment of Cheese Milk. *Foods* **2021**, *10*, 184. https://doi.org/10.3390/foods10010184

Received: 11 December 2020
Accepted: 15 January 2021
Published: 18 January 2021

Publisher's Note: MDPI stays neutral with regard to jurisdictional claims in published maps and institutional affiliations.

Copyright: © 2021 by the authors. Licensee MDPI, Basel, Switzerland. This article is an open access article distributed under the terms and conditions of the Creative Commons Attribution (CC BY) license (https://creativecommons.org/licenses/by/4.0/).

1. Introduction

Fresh, spreadable cheese varieties with paste-like consistency such as Quark-type and Cream cheese come mainly from combined acid-rennet curds, in which hydrated acid casein gel particles are dispersed in whey [1,2]. According to Codex Alimentarius [3], cheeses of the category "Cream cheese" are for the most part intended for direct consumption, can spread and mix readily with other foods, and contain at least 22% dry matter and 67% moisture on a fat-free basis. A brief literature outline of the effects and mechanisms observed in combined acid–rennet curd/cheese due to interventions on cheese milk such as fat reduction, heating, and homogenization, which are related to present study, is given below.

To counteract the effects of fat reduction on the paracasein matrix and on the subsequent ripening course, modifications are applied in (i) the treatment of reduced-fat cheese milk, i.e., heating conditions, homogenization, fat mimetic,s or replacers, and (ii) the cheese-making conditions, i.e., starter mixture, curdling, cutting, scalding, in-vat acidification, salting [4,5]. Milk heat treatment can be a means to control the texture and yield of dairy products, e.g., heating at 95 °C for 15 min or at 90 °C for 10 min increases Quark yield, moisture, and firmness [6,7]. Treatment more intense than typical, low pasteurization causes an in situ denaturation of whey proteins that form soluble aggregates and complexes with casein micelles, which in turn reduce syneresis and moisture loss from the curd [8–10].

The complexation of casein with whey proteins denatured by the heating of skim milk at 80 °C for 30 min hinders the rennet action on κ-casein and the concomitant aggregation of rennet-altered micelles. Therefore, smaller pores, lower permeability, lower whey separation, less sensitivity to large-scale arrangements, and higher firmness are observed in the resultant combined acid–rennet gels compared to unheated counterparts [11]. Whey proteins denatured by the heating of milk at 82 °C and 90 °C for 5 min participate in the gel formation of Quark. As a result, a regular microstructure with large protein aggregates, branched clusters, and numerous small pores is developed opposite to the larger pores of pasteurized milk at 72 °C for 16 s [12]. Similarly, an increase of pasteurization temperature from 65 to 75 °C for 30 min increased substantially the moisture and yield of a spreadable goat milk [13]. However, it has been reported that the increase of milk heating temperature from 80 to 90 °C for 5 min has no significant effect on syneresis and water-holding capacity and decreases the firmness and stickiness of Quark-type gels [14].

The homogenization of cheese milk or cream under specific conditions has been proposed as a means to increase the moisture and improve the texture and heat-induced functionality of reduced- or low-fat cheeses [4,9]. The reduced size of homogenized milk fat globules (MFG) and the casein absorption onto their surface enables them behave as pseudo-protein particles interacting with casein micelles during coagulation opposite to the MFG of the unhomogenized milk, which are embedded into the paracasein matrix [15,16]. Consequently, the fusion of the paracasein micelles and whey removal are not favored, resulting in reduced gel strength and poor syneresis, high moisture, and increased fineness of the homogenized milk curds that can deviate the ripening course and functionality [4,9]. The homogenization of cheese milk is applied for the manufacture of soft, high-fat, acid-rennet curd Quark or Cream cheese. It counteracts creaming during the long curdling period, reduces fat loss in the whey, and increases the effective protein concentration. Moreover, the participation of homogenized MFG in gel formation contributes to the development of a homogeneously soft and smooth texture on subsequent acidification without wheying off or drying [1,4]. In brief, the following mechanisms have been suggested [17,18]. Homogenization at 20 MPa, at normal milk pH 6.7, decreases the ζ-potential of new homogenized MFG that becomes similar to that of casein micelles that are absorbed onto their surface. The protonation of phosphate and carboxylic residues due to acidification decreases further the ζ-potential of both new MFG and caseins to -2.3 and -4.3 mV, respectively whereas in unhomogenized milk, it is -6.6 mV. Furthermore, the replacement of stabilizing highly structured glycocalyx of the native milk fat globule membrane (MFGM) by the more flexible casein and the decrease of electrostatic repulsions at pH 4.5 enhance the hydrophobic interactions between new MFG and casein micelles. Since casein micelles are present at both sides of interaction, homogenized MFG become active fillers [18]. Heating corresponding to an 80% denaturation of total whey proteins reduces further the ζ-potential of the casein micelles and homogenized MFG because denatured whey proteins are complexed onto the casein surface or co-absorbed with casein onto the homogenized MFGM [17]. Moreover, additional strong hydrophobic interactions between denatured whey proteins result in stranded protein structures and in stiffer and more connected acid gel compared to their unheated homogenized counterpart [17].

The effects of milk homogenization or fat reduction on the physical properties of Cream cheese have been reported. The increase of homogenization pressure from 0 to 25 MPa of milk with protein-to-fat (P/F) ratio 0.97 can decrease approximately five times the $D_{4,3}$ and ten times the $D_{3,2}$ descriptors of particles and increase significantly the firmness, spreadability, and viscosity of medium-fat Cream cheese; at 100 MPa, no further significant differentiation is observed [19]. It has been shown that homogenization at 30 MPa destabilizes acidified whey protein water/oil emulsions when there are not enough non-denatured whey proteins to cover the interface [20]. Homogenization at 15 MPa decrease by eight times the $D_{4,3}$ and more than ten times the $D_{3,2}$ of particles in Cream cheese with a P/F ratio of 0.33, resulting in higher consistency or stiffening of the gel network. Interestingly, the protein content of the bulk phase not absorbed onto the homogenized

MFG decreased by seven times [21]. The additional whey protein denaturation caused by outlet heating of post-homogenization of an acidified Cream cheese model enhanced protein-mediated interactions between fat globules, resulting in higher firmness and lower spreadability [22]. The higher hardness, elasticity cohesiveness, and spreadability of full-fat compared to low-fat commercial Cream cheeses have been assigned to the high numbers of homogenized MFGs that counteract the higher protein content of the latter [23]. The decrease of milk fat results in larger particles in the cheese matrix; i.e., the mean particle size in Cream cheese with low-0.5%, medium-5.5%, and high-11.6% fat content has been found at 18.2, 11.3, and 8.1 μm, respectively [24]. The physical properties of commercial Cream cheeses with different fat contents are largely shaped by the effect of stabilizers on their microstructure; however, low fat content and high moisture are consistent with less structured and more spreadable products [25].

Soft, acid cheeses with paste-like consistency are manufactured in various regions of Greece from cow, ewe, goat milk, or their mixtures. They result from the rennet and/or acid coagulation of rather high heat-treated, full-fat, unhomogenized milk. The next processing steps are acidification/curdling, draining by means of cloth filter bags, gentle homogenization by hand, and often salting of the drained curd and a very short or no ripening period; no stabilizers or other additives are used. Differences in the cheesemaking protocols shape a variety of spreadable cheeses, with high moisture (60–75%) and fat on a dry basis (40–55%), rather low salt (1–1.8%) and pH $\cong$ 4.5; some of them are protected denomination of origin (PDO) cheeses, i.e., Galotyri, Katiki, Piktogalo Chanion, Anevato, Kopanisti, and Xygalo Siteias [26–30].

Considering the consumers' interest in this cheese type and the current demand for reduced-fat and reduced-energy dairy products, the present study was undertaken to investigate the effect of homogenization and fat reduction in combination with variable heat treatments of cow milk on the characteristics of Quark-type cheese.

2. Materials and Methods

2.1. Milk Treatments and Analyses

Cheese milk homogenization and continuous heat treatments by means of tubular heat exchanger were carried out in the laboratory heating system HT220 HTST/UHT (OMVE Lab & Pilot Equipment, 3454 MZ De Meern, The Netherlands) equipped with an in-line two-stage homogenizer (Twin Panda, Gea Niro Soavi, Type NS2002H). In one experimental day, raw cow milk from the premises of Agricultural University of Athens was divided in two parts. The first part (milk UH) was heated in the laboratory heating system at 68 °C (1), 73 °C (2), 78 °C (3), 85 °C (4), and 100 °C (5) for 16 s and cooled down at 30 °C. In addition, a portion of milk was heated under batch conditions in an open vat at 90 °C for 5 min (6) and then cooled immediately in an ice bath. The second part was similarly heat treated (conditions 1–6) but after in-line two-stage homogenization at 25 and 5 MPa (milk H). The next day, the same treatments were applied in reduced-fat (RF) milk that was a mixture of raw full-fat (FF) milk and milk skimmed by means of a cream separator. The protein-to-fat (P/F) ratio of RF milk increased to 1.62 from 0.95 of FF, which corresponded to a 38% fat reduction. In each experimental day, 12 milk portions were prepared and along with the two controls (H and UH, not heat-treated) resulted in 14 milk samples. The two-day set of experiments was performed in triplicate.

The gross composition of milk samples was estimated by Milkoscan-FT120 134 (Foss, Hilleroed, Denmark), titratable acidity by means of the Dornic method and pH by means of a pHmeter (WMW Multiline Multi 3420, Fisher Scientific, Loughborough, Leicestershire, UK). Assessment of the heat denaturation of major whey proteins α-lactalbumin (α-la) and β-lactoglobulin (β-lg) was based on their residual content in the pH 4.6 soluble fraction of milk, which was analyzed by the RP-HPLC method, as described by Sakkas et al. [31].

2.2. Cheesemakings

Based on the results for whey protein denaturation, four of the above-mentioned heat treatments—that is, 2, 4, 5, and 6—were selected for the treatment of cheesemilk utilized for the manufacture of Quark-type combined curd cheeses. In one experimental day, unhomogenized (UH) and homogenized (H) FF milk portions were heated at 73 °C (2), 85 °C (4), and 100 °C (5) for 16 s and at 90 °C for 5 min and then cooled down to 30 °C as previously described. Ten kg milk from each homogenization–heating combination were collected in a sterilized container, and at 25 °C, 1.8% dry salt was added. Subsequently, inoculation by means of a commercial starter took place (DVS MO-10-Chr. Hansen Holding A/S, Hoershom, 94 Denmark). Then, calf rennet powder with clotting activity 1:100,000 was added at a ratio of 10 mg per L of milk. After gentle stirring, the inoculated milk was left for approximately 18 h at 18–20 °C for acidification to pH 4.6 and curd formation. The curd cut into 2 × 2 × 2 cm cubes was transferred into cloth filter bags for draining at 18–20 °C for approximately 20 h. Then, the fresh cheese was weighed for the estimation of yield, homogenized gently by hand, packed in sterilized containers, and stored at 4 °C. Within the same week, the same experimental cheese makings were carried out, using RF cheese milk.

In each experimental day, 8 cheese makings were carried out. The FF-RF set of experimental cheese makings was performed in triplicate.

2.3. Cheese Analyses

Cheeses were analyzed on 5 and 20 day of storage; in total, 96 cheese samples were analysed.

Gross composition of cheeses was estimated by FoodScan-Dairy Near Infrared (NIR) analyzer (Foss, Hilleroed, Denmark). Cheese pH was determined in a dispersion of cheese in water at 1:1 ratio by means of a pH meter. Cheese titratable acidity was estimated in a filtrate of a cheese dispersion; ten g cheese dispersed in 92 mL of distilled water were filtered with coarse filter paper, and the acidity of 15 mL of the filtrate was determined using N/10 NaOH and phenolphalein.

Residual sugars and organic acids, i.e., lactose, glucose, galactose, citric acid, and lactic acid were determined by HPLC by means of the Aminex HPX-87H column (Biorad Inc., Hercules, CA, USA) [32,33]. Twenty g of cheese were diluted to 100 mL with Biggs-Szijarto solution i.e., 25 g zinc acetate dihydrate, 12.5 g phosphotungstic acid monohydrate, and 20 mL glacial acetic acid in 200 mL. After mixing and standing for 10 min, the cheese dilution was filtered with Whatman No. 1 filter paper. One mL of filtrate was mixed with 100 µL perchloric acid (70%) and after an overnight stay at 4 °C, the mixture was centrifuged at 10,000× g for 30 min at 4 °C. The supernatant was filtered with 0.45 µm before HPLC analysis; the quantification was based on the respective standard curves.

Texture profile analysis (TPA) was carried out in duplicate at 15 °C using a 5 kg load cell and a 35 mm plunger by means of the double bite test [34,35] using a Shimadzu Testing Instrument AGS-500 NG (Shimadzu Corporation, Kyoto, Japan).

The nitrogen (N) content of cheese (total N, TN), the water-soluble nitrogen (WSN) and the nitrogen soluble in 12% trichloroacetic acid (TCAN) were determined by the Kjeldahl method, as described by Moschopoulou et al. [36]. In brief, 30 g of cheese were homogenized with 90 g of distilled water using Ultra Turrax (IKA Works, Inc., Wilmington, NC, USA) at 9500 rpm for 2 min. After 60 min at 40 °C, the cheese dispersion was re-homogenized under the same conditions followed by centrifugation at 3000× g, for 30 min at 6 °C. The supernatant filtered through Whatman No. 1 filter paper was the WSN fraction. For the preparation of TCAN fraction, 75 mL of WSN were mixed with an equal volume of 24% (w/v) trichloroacetic acid (TCA), held overnight at 4 °C, and then filtered through Whatman No. 1 filter paper.

Organoleptic evaluation of cheeses at 20 day of storage was performed by a panel of four experienced laboratory staff members. Cheeses coded by three-digit code were randomly presented in the panel for the evaluation of appearance, texture, and flavor using a scale from 0 to 10 points. The total organoleptic score—expressed as percentage—

was the sum of the appearance, texture, and flavor scores multiplied by 1, 4, and 5, respectively. Moreover, the panel members noted the existence of specific defects, such as non-homogenous color, spoiled surface, dryness, stickiness, lack of spreadability, excessive acidity and saltiness, bitterness, metallic taste, or yeast flavor.

2.4. Statistical Analysis

The effects of homogenization, fat reduction, heat treatments, and their interactions on milk and cheese characteristics were assessed by multifactor ANOVA, whereas the least significance (LSD) method ($p < 0.05$) was applied to test the significant differences. Analysis was performed by means of Statgraphics Centurion XVI (Manugistics, Inc., Rockville, MA 20852, USA).

3. Results and Discussion

3.1. Characteristics of Cheese Milk

The results of milk analyses are presented in Table 1, which are grouped by the three experimental factors. No significant effects of the interactions were observed, with the exception of residual α-lactalbumin (α-la), which was affected by the interaction of fat content and heat treatment. According to multifactor ANOVA, the effect of homogenization on the compositional parameters and residual whey proteins of cheese milk was limited to acidity, which can be assigned to the required additional time and treatments. As expected, the fat reduction statistically significantly ($p < 0.05$) affected the compositional parameters of cheese milk. In fact, the protein-to-fat ratio increased from 0.95 in full-fat (FF) to 1.62 in reduced-fat (RF) milk. It is evident (Table 1) that the significant effect ($p < 0.05$) of the heat treatment on the compositional parameters of milk was due to the significantly different ($p < 0.05$) milk 6, which was heated under batch conditions in a more or less open container at 90 °C for the rather long period of 5 min; under such conditions, a limited evaporation can occur. The increase of heat load significantly ($p < 0.05$) reduced the native major whey protein content of milk. It is well established that α-la is less heat-sensitive than β-lg; for example, native β-lg was not found at temperatures >100 °C, whereas 43 mg/L residual α-la was estimated at 130 °C [31]. The heating conditions of milk 6 denatured fully β-lg and 75% of α-la. Moreover, the pH of milk 6 was significantly lower ($p < 0.05$), which was apparently due to the equilibria change of calcium phosphate between casein micelles and milk serum [37].

Table 1. Means of compositional parameters and residual native whey proteins of cheese milk used in the experiments.

Factors	n	Fat %	Protein %	Lactose %	TS %	α-la %	β-lg %	pH	Acidity %
Homogenization									
H	42	2.65	3.21	4.37	11.34	79.52	63.54	6.76	0.13 [b]
UH	42	2.67	3.2	4.37	11.34	77.86	62.99	6.74	0.12 [a]
SE		0.02	0.01	0.01	0.03	0.52	1.18	0.01	0.001
LSD		0.06	0.03	0.03	0.1	1.90	3.37	0.03	0.003
Fat content									
FF	42	3.30 [b]	3.13 [a]	4.40 [b]	11.92 [b]	77.88	63.62	6.69 [a]	0.13 [b]
RF	42	2.03 [a]	3.28 [b]	4.34 [a]	10.76 [a]	79.49	62.91	6.81 [b]	0.12 [a]
SE		0.02	0.01	0.01	0.03	0.52	1.18	0.01	0.001
LSD		0.06	0.03	0.03	0.1	1.89	3.37	0.03	0.003
Heat treatment									
NHT	12	2.61 [a]	3.20 [a]	4.33 [ab]	11.25 [a]			6.75 [ab]	0.13
1 (68 °C/16 s)	12	2.60 [a]	3.14 [a]	4.29 [a]	11.13 [a]	98.60 [e]	99.52 [e]	6.77 [b]	0.12
2 (73 °C/16 s)	12	2.65 [a]	3.17 [a]	4.33 [ab]	11.26 [a]	95.98 [e]	99.59 [e]	6.77 [b]	0.13
3 (78 °C/16 s)	12	2.65 [a]	3.18 [a]	4.34 [ab]	11.28 [a]	92.57 [d]	90.12 [d]	6.77 [b]	0.13
4 (85 °C/16 s)	12	2.66 [a]	3.19 [a]	4.35 [b]	11.31 [a]	84.98 [c]	67.82 [c]	6.75 [ab]	0.12
5 (100 °C/16 s)	12	2.66 [a]	3.18 [a]	4.35 [b]	11.30 [a]	74.51 [b]	20.74 [b]	6.74 [ab]	0.13

Table 1. *Cont.*

Factors	n	Fat %	Protein %	Lactose %	TS %	α-la %	β-lg %	pH	Acidity %
6 (90 °C/5 min)	12	2.80 [b]	3.36 [b]	4.59 [c]	11.84 [b]	25.49 [a]	1.82 [a]	6.70 [a]	0.13
SE		0.04	0.02	0.02	0.07	0.90	2.01	0.02	0.002
LSD		0.04	0.06	0.06	0.06	3.28	5.84	0.02	0.02

Different letters indicate statistically significant differences (least significant difference (LSD), $p < 0.05$) in the group of means assigned to each experimental factor. SE, standard error; H, homogenized; UH, unhomogenized; FF, full fat; RF, reduced fat; TS, total solids; α-la and β-lg, residual α-lactalbumin and β-lactoglobulin; NHT, not heat-treated.

3.2. Compositional Characteristics of Cheese

Selected heat treatments 2, 4, 5, and 6 with variable denaturation level of whey proteins were applied in homogenized and unhomogenized full- and reduced-fat cheese milk with the aim of assessing their impact on physicochemical characteristics of the resultant cheeses. Batch heating at 90 °C for 5 min, corresponding to cheese 6 in the present experiments is close to the traditional practice and is a typical treatment for the manufacture of this cheese type or yoghurt. The mean composition of the experimental cheeses, grouped according to the experimental factors, is shown in Table 2. All the experimental factors affected the composition of the experimental cheeses, except for storage. In fact, storage caused an increase ($p < 0.05$) of acidity expressed as percentage of lactic acid that can be attributed to the catabolism of residual lactose by the starters.

Table 2. Means of compositional parameters of Quark-type cheeses made from unhomogenized and homogenized full- and reduced-fat milk heated under different conditions.

Factors	n	Protein %	Moisture %	Salt %	Fat %	pH	Acidity %	Yield %	P/F
Homogenization									
H	48	10.97 [a]	74.14 [b]	1.37	11.4 [a]	4.27	0.58	0.27 [b]	0.96
UH	48	12.18 [b]	70.90 [a]	1.34	12.94 [b]	4.26	0.59	0.24 [a]	0.94
SE		0.07	0.12	0.01	0.83	0.009	0.007	0.003	
LSD		0.205	0.363	0.044	0.236	0.027	0.022	0.008	
Fat content									
FF	48	10.31 [a]	71.96 [a]	1.26 [a]	13.95 [b]	4.28 [b]	0.56 [a]	0.28 [b]	0.74
RF	48	12.84 [b]	73.08 [b]	1.45 [b]	10.39 [a]	4.24 [a]	0.61 [b]	0.24 [a]	1.24
SE		0.07	0.12	0.01	0.08	0.009	0.007	0.003	
LSD		0.205	0.363	0.044	0.236	0.027	0.022	0.008	
Heat treatment									
2 (73 °C/16 s)	24	11.84 [b]	71.40 [a]	1.36	12.92 [c]	4.26	0.61 [b]	0.23 [a]	0.92
4 (85 °C/16 s)	24	11.62 [ab]	72.29 [b]	1.35	12.51 [b]	4.24	0.58 [ab]	0.25 [b]	0.93
5 (100 °C/16 s)	24	11.36 [a]	73.28 [c]	1.33	11.79 [a]	4.25	0.58 [ab]	0.26 [c]	0.96
6 (90 °C/5 min)	24	11.46 [a]	73.11 [c]	1.38	11.47 [a]	4.28	0.58 [a]	0.30 [d]	1.00
SE		0.10	0.18	0.02	0.11	0.01	0.01	0.004	
LSD		0.290		0.063	0.333	0.039	0.031	0.012	
Storage									
5 day	48	11.56	72.51	1.36	12.17	4.26	0.57 [a]		0.95
20 day	48	11.59	72.53	1.35	12.17	4.26	0.61 [b]		0.95
SE		0.07	0.12	0.01	0.08	0.009	0.007		
LSD		0.205	0.363	0.044	0.236	0.027	0.022		

Different letters indicate statistically significant differences (LSD, $p < 0.05$) in the group of means assigned to each experimental factor. SE, standard error; H, homogenized; UH, unhomogenized; FF, full fat; RF, reduced fat.

Apparently, the fat reduction of cheese milk significantly ($p < 0.05$) affected the parameters of Table 2. On average, full-fat (FF) cheeses contained 71.96% moisture, 13.95% fat, 10.31% protein, and 1.26% salt. The respective contents for the reduced-fat (RF) counterparts were 73.08%, 12.84%, 1.45%, and 10.39%. The protein-to-fat ratios were lower than those of the respective cheese milks, i.e., 0.74 and 1.24 for FF and RF cheeses indicating higher protein loss than fat in the whey, during draining.

Homogenization significantly ($p < 0.05$) improved the moisture retention in cheese curd, and consequently, fat and protein contents were reduced. On average, homogenization increased the retention of protein in cheese; the P/F ratio was 0.96, which was higher than the 0.94 of its unhomogenized counterpart. Indeed, the mean protein on dry matter of H cheeses was higher, i.e., 42.42% compared to 41.86% of UH cheese. The mechanisms related to the effect of homogenization with emphasis on acid gels have been concisely presented in the interaction. Opposite to the native, the homogenized MFGs participate in the casein matrix with the interactions being stronger at acidic pH [15,16,18] such as the pH of the cheeses of the present study. These interactions do not favor the fusion of casein micelles; consequently, syneresis is impaired, and cheese moisture increases. The effect of homogenized MFGs on the paracasein matrix has been used as a means to increase the moisture and functionality of rennet-clotted semihard reduced and low-fat cheeses [4]. As mentioned in the Introduction, the homogenization of Cream cheese milk induces an intense decrease of the size of particles within the cheese matrix and reduces the protein amount that is not absorbed onto the homogenized MFGs [19,21].

The importance of milk heat treatment for this type of cheese from combined acid curds formed by the simultaneous acidity development and rennet action on casein has been also highlighted in the Introduction. The participation of denatured whey proteins in the gel and the hindering of rennet action on the casein micelle–whey protein complexes result in curds with smaller pores and lower whey separation [11,12] and finally in spreadable cheese with higher moisture e.g., [13]. A statistically significant increase of cheese moisture and yield caused by a high heat treatment of Quark cheese milk at 90 °C for 10 min compared to typical pasteurization has been reported by Kelly and O'Donelly [7]. They suggested that the level of denaturation of α-la is associated to rheology and texture, while the level of denaturation of β-lg affects moisture and yield. Miloradovic et al. [14] found that the composition of Quark-type cheeses manufactured from cow or goat milk were not affected by the increase of heat treatment of milk from 80 to 90 °C for 5 min, but they report an average increase of yield from 27.4 to 31.3%, respectively, for cow milk cheese. The increase of the intensity of heat treatment (Table 2) improved the P/F ratio apparently due to the inclusion of heat-denatured whey proteins into the curd. The P/F ratio of cheese from the most severely heated cheese 6 was equal to 1, while in cheese 2 from pasteurized milk, it was 0.92. Interestingly, the experimental cheeses made from high heat-treated milk 5 and 6 did not differ in respect to moisture, fat, and protein content, despite the significant differences ($p < 0.05$) in the denaturation level of whey proteins (Table 2).

Significant effects ($p < 0.05$) were observed for the interactions of homogenization–heating on moisture, fat, and protein. The moisture on non-fat substances (MNFS) of cheeses made from homogenized milk (H) was higher than those made from unhomogenized milk (UH). In particular, it was >2.4% higher for the three continuous-type heat treatments 2, 4, and 5 (Figure 1). However, the MNFS difference between H and UH cheese was much lower (1.1%) when the most intense batch treatment at 90 °C for 5 min was applied, indicating that the effect of homogenization is less pronounced when severe denaturation of whey protein takes place. Similarly, the P/F ratio was higher in H cheeses than in UH when they were treated by the continuous heating method, but no difference was observed for the more severe treatment 6 (Figure 1).

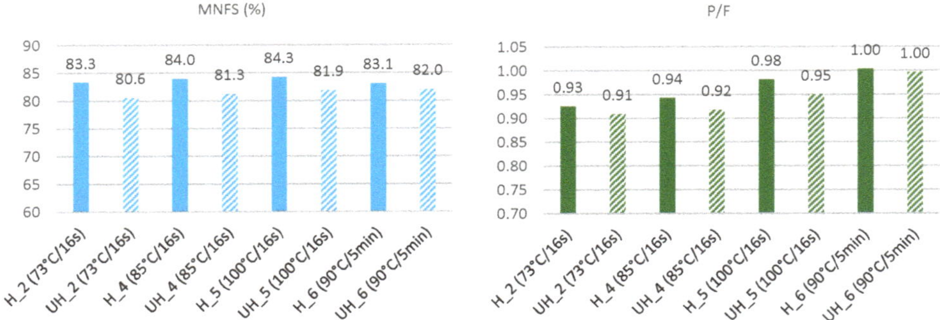

Figure 1. Moisture on non-fat substances (MNFS, %) and protein-to-fat ratio (P/F) of Quark-type cheeses made from homogenized (H) and unhomogenized (UH) milk heat-treated under different conditions.

Significant ($p < 0.05$) effects of the interaction of homogenization–fat content on cheese moisture and fat were observed. The effect of homogenization on moisture retention was more pronounced in full-fat cheeses. The mean moisture contents of H and UH FF cheeses were 74.08% and 69.84%, and the respective contents for RF cheeses were 74.21% and 71.95%. In particular, homogenization increased the MNFS of FF cheese by 2.7% and that of RF cheese by 1.8%.

3.3. Residual Sugars and Organic Acids of Cheese

The concentrations of lactose, monosaccharides, and organic acids were not affected significantly by milk homogenization (Table 3). They were affected by the fat content being significantly higher ($p < 0.05$) in RF cheeses due to the reduced participation of fat into total solids (Table 1). In addition, lactic acid concentration of RF cheese coincided with their significantly ($p < 0.05$) lower pH and higher titratable acidity (Table 1). Heat treatment 6 at 90 °C for 5 min did not affect the monosaccharides and lactic acid but resulted in higher ($p < 0.05$) residual lactose and citric acid contents. The latter could imply a retardation of starter activity, which is not desirable in cheesemaking. During storage, there was a statistically significant ($p < 0.05$) reduction of galactose indicating the slow catabolism from cheese microflora opposite to glucose. The significant ($p < 0.05$) increase of citric acid during storage can be assigned to changes of the paracasein matrix, i.e., proteolysis in Section 3.5—that under the low pH of cheese results in the solubilization of citrate associated with colloidal calcium phosphate [37].

Table 3. Means of residual sugars and organic acids (g per 100 g cheese) of Quark-type cheeses made from unhomogenized and homogenized full-fat and reduced-fat milk heated under different conditions.

Factors	Lactose	Glucose	Galactose	Citric Acid	Lactic Acid
Homogenization					
H	3.90	0.177	0.399	0.354	0.833
UH	4.00	0.197	0.493	0.357	0.923
SE	0.043	0.009	0.035	0.086	0.033
LSD	0.119	0.024	0.099	0.024	0.095
Fat content					
FF	3.87 [a]	0.173 [a]	0.365 [a]	0.353	0.720 [a]
RF	4.03 [b]	0.200 [b]	0.527 [b]	0.358	1.036 [b]
SE	0.043	0.009	0.035	0.086	0.032
LSD	0.119	0.027	0.099	0.024	0.095

Table 3. *Cont.*

Factors	Lactose	Glucose	Galactose	Citric Acid	Lactic Acid
Heat treatment					
2 (73 °C/16 s)	3.82 [a]	0.191	0.436	0.331 [a]	0.859
4 (85 °C/16 s)	3.96 [a]	0.193	0.484	0.361 [a,b]	0.910
5 (100 °C/16 s)	3.86 [a]	0.180	0.429	0.360 [a,b]	0.884
6 (90 °C/5 min)	4.15 b	0.183	0.434	0.369 [b]	0.859
SE	0.059	0.013	0.490	0.012	0.048
LSD	0.170	0.035	0.140	0.035	0.134
Storage					
5 day	3.91	0.189	0.502 [b]	0.331 [a]	0.883
20 day	3.98	0.185	0.389 [a]	0.379 [b]	0.872
SE	0.043	0.009	0.035	0.085	0.034
LSD	0.120	0.025	0.099	0.023	0.095

Different letters indicate statistically significant differences (LSD, $p < 0.05$) in the group of means assigned to each experimental factor. SE, standard error; H, homogenized; UH, unhomogenized; FF, full fat; RF, reduced fat.

The lactose content of cheese depends for the most part on the cheesemaking conditions that determine its residual amount in the curd. Then, the fermentation of residual lactose and of its subunits glucose and galactose takes place by starters in the first hours or days after manufacture. Moreover, citrate fermentation by particular lactoccoci strains can occur in particular cheese varieties. Pathways of lactose, lactic acid, and citrate catabolism in cheese are presented in detail by McSweeney et al. [38]. The cheesemaking conditions of the present study did not favor the removal of a large quantity of whey, and therefore, a high residual lactose content remained in the cheese mass. Moreover, only a part of it can be catabolized by the starters due to the inhibitory effect of the produced lactic acid on their activity. On average, FF cheeses contained 3.86 ± 0.346 % lactose and 0.720 ± 0.117% lactic acid, whereas the respective contents of RF counterparts were 4.02 ± 0.274 and 1.035 ± 0.310%, which are similar with those reported for this category of cheese. The sum of lactose and lactic acid of skim milk Quark with <1.8% fat and single Cream cheese with 19.5% fat is 3–4% and 3.5%, respectively [1]. Lower contents, i.e., 2.27% lactose and 0.33% lactic acid, reported for Cream cheese with 33.6% fat, can be due to the marked participation of fat in the total solids [1]. A typical gross composition presented by Farkye [2] indicates that low-fat Quark with 82% moisture and pH 4.4–4.6 contains 3–4% lactose, while its full-fat counterpart with 76% moisture and pH 4.5–4.6 contains 2.5–3.5% lactose. A high variability in the lactose content of various artisanal fresh acid-coagulated cheeses with moisture content 44.3–51.3% has been pointed out by Zeppa and Rolle [39]. The lactose content ranged from 2.4 to 22.8 mg per g cheese, the lactic acid ranged from 6 to 27.85 mg/g, glucose was absent, galactose ranged from 0.006 to 4.07 mg/g, and citric acid was from 0.37 to 0.88 mg/g. Papadakis and Polychroniadou [40] reported 14–18 mg lactic acid and 0.12–0.88 mg citric acid per g feta cheese; the respective amounts per g sheep milk yoghurt were 13–16 mg and 1.4–2.3 mg.

3.4. Texture Profile Analysis of Cheese

The mean parameters and statistical analysis of texture profile analysis (TPA) of Quark-type experimental cheeses are presented in Table 4 grouped by the experimental factors. Storage exhibited no significant effect ($p > 0.05$) on the texture of cheeses, while the other factors significantly affected ($p < 0.05$) most of these parameters. Moreover, a significant ($p < 0.05$) effect of the interaction homogenization–heating on the hardness was observed.

Table 4. Means of the parameters of texture profile analysis of Quark-type cheeses made from unhomogenized and homogenized full- and reduced-fat milk heated under different conditions.

Factors	Hardness	Adhesiveness	Springiness	Cohesiveness	Gumminess	Chewiness
Homogenization						
H	13.16 [a]	−64.08 [b]	1.02	0.5	6.52 [a]	6.70 [a]
UH	16.41 [b]	−81.58 [a]	1.00	0.51	8.34 [b]	8.42 [b]
SE	0.34	2.46	0.009	0.007	0.17	0.19
LSD	0.984	6.92	0.026	0.02	0.495	0.536
Fat content						
FF	13.69 [a]	−68.04 [b]	1.01	0.51 [b]	7.01 [a]	7.11 [a]
RF	15.88 [b]	−77.62 [a]	1.02	0.49 [a]	7.84 [b]	8.01 [b]
SE	0.34	2.46	0.008	0.007	0.17	0.19
LSD	0.984	6.92	0.026	0.02	0.495	0.536
Heat treatment						
2 (73 °C/16 s)	13.84 [a]	−66.62 [a]	0.99 [a]	0.51 [a,b]	6.95	6.90 [a]
4 (85 °C/16 s)	14.86 [a,b]	−78.16 [b]	1.03 [b]	0.51 [a,b]	7.55	7.84 [b]
5 (100 °C/16 s)	14.53 [a,b]	−72.39 [a,b]	1.01 [a,b]	0.52 [b]	7.57	7.67 [b]
6 (90 °C/5 min)	15.91 [b]	−74.15 [a,b]	1.02 [b]	0.48 [a]	7.64	7.83 [b]
SE	0.49	3.48	0.01	0.01	0.24	0.26
LSD	1.39	9.79	0.037	0.028	0.7	0.758
Storage						
5 day	14.64	−74.09	1.01	0.52	7.48	7.62
20 day	14.93	−71.06	1.01	0.50	7.37	7.50
SE	0.34	2.46	0.009	0.07	0.17	0.19
LSD	0.984	6.92	0.026	0.02	0.495	0.536

Different letters indicate statistically significant differences (LSD, $p < 0.05$) in the group of means assigned to each experimental factor. SE, standard error; H, homogenized; UH, unhomogenized; FF, full fat; RF, reduced fat.

Gunasekaran and Ak [35] described the TPA parameters as follows. Hardness is the force necessary to attain given information. Adhesiveness is the work necessary to overcome the attractive forces between the surface of the food and surface of other materials with which the food comes into contact. Springiness is the distance recovered by the sample during the time between end of the first bite and start of the second bite. Cohesiveness is the strength of the internal bonds making up the body of the product. Gumminess is the energy needed to disintegrate a semisolid food until it is ready for swallowing. Chewiness is the energy needed to chew a solid food until it is ready for swallowing.

The discussion of our TPA findings in relation to published studies for other paste-like spreadable cheeses is not always adequate, because they are mostly Cream cheese varieties with a rather low protein-to-fat ratio. Quark-type cheeses made from alternative raw materials of dairy origin such as buttermilk [41,42] are not included in the discussion due to the particularities of raw materials that can differentiate the formation and properties of curd.

According to Table 4, the fat reduction resulted in cheeses (RF) with significantly ($p < 0.05$) higher hardness, gumminess, and chewiness and significantly lower adhesiveness and cohesiveness compared to the full-fat counterpart. The effects of fat reduction on the texture profile of rennet-curd cheeses have been reviewed [5,35]. In brief, low-fat rennet-curd cheeses have a more close structure that increases hardness, dryness, graininess, and springiness and decreases adhesiveness and cohesiveness. Moisture increase has the opposite effect on hardness and springiness. However, cheeses of the present experiment resulted from combined curds formed at low pH. Differences in the mechanisms that take place during the formation of cheese curd could affect some textural properties in a different manner, e.g., fat reduction by 10–30% in model cheese analogues based on acid–casein coagulation increased significantly the hardness and adhesiveness [43]. Nguyen et al. [24] report an increase of hardness and adhesiveness when the fat content of Cream cheese is reduced to reach a P/F ratio 0.97, which is much lower than that of the present RF cheeses

(Table 2). As presented in the Introduction, the authors attributed these differences to the estimated increase of larger particles in a medium- and low-fat matrix. In addition, they assigned the observed higher stickiness—which is related to adhesiveness—to the increase of protein content due to fat reduction. Moreover, the comparison of the present findings with studies on commercial cheeses is not adequate due to the use of stabilizers in the latter. Stabilizers affect along with fat, moisture, and microstructure the physical properties of this type of cheese [25]; for example, Kealy [44] estimated a lower hardness and adhesiveness and higher cohesiveness for commercial Cream cheeses with stabilizers compared to their full-fat counterparts.

The results of Table 4 for the effect of various heat treatments are in accordance to the brief overview in the Introduction. The heating of cheese milk of combined gels at conditions that induce the denaturation of whey proteins and therefore their complexation with casein micelles increases firmness by decreasing the gel pores, impairing whey separation and improving the regularity of the microstructure [11,12]. The increase of firmness of Quark cheese by heat treatments higher than pasteurization has been reported [6,7]. On the other hand, the lower firmness of Quark-type cheese manufactured from milk batch-heated at 90 °C compared to 80 °C for its 5 min counterpart has been reported [14]; apparently, the batch heating conditions should be taken into consideration for this effect.

The optimum heat treatment for combined curds induces a whey protein denaturation level of 75–80% [1]. Therefore, according to the denaturation level shown in Table 1, batch treatment at 90 °C for 5 min is not adequate for this curd type. The extended denaturation of whey proteins observed in cheese milk 6 (Table 1) resulted in the most hard and springy and the least cohesive cheese ($p < 0.05$). On the other hand, cheese from pasteurized milk was the most soft and adhesive and the least springy ($p < 0.05$). Finally, statistically significant differences ($p < 0.05$) between TPA parameters of heat treatments 4 (85 °C/16 s) and 5 (100 °C/16 s) were scarce. Therefore, the texture of cheeses was not significantly differentiated by heat treatment of milk that induced the denaturation of β-lg and α-la by (i) 37% and 15%, respectively and (ii) by 80% and 25%, respectively.

The mechanisms involved in the formation of acid curd from homogenized milk have been described in the first section of the present article; in conclusion, homogenized MFG acts as active fillers in the acidified casein network [16,17]. The considerable decrease of particles in Cream cheese induced by milk homogenization at conditions similar to ours result in higher firmness, consistency, and spreadability [19,21]. Coutouly et al. [22] report that the post-homogenization of an acidified Cream cheese model resulted in more firm and less spreadable cheese. On the other hand, Brighenti et al. [21] observed that the effect of homogenization counteracted the effect of fat reduction/protein increase in Cream cheese, and increased the hardness, elasticity, cohesiveness, and spreadability of its full-fat counterpart. As shown in Table 4, cheeses from homogenized milk were significantly ($p < 0.05$) softer with significantly lower adhesiveness, gumminess, and chewiness than counterparts from unhomogenized milk. Therefore, cheeses H were less sticky and more easily masticated.

Statistical analysis showed that there was a significant ($p < 0.05$) interaction between homogenization and heat treatment, as presented in Figure 2. Similarly, to the above-mentioned weaker effect on the MNFS of cheeses 6, the homogenization of milk did not exhibit any significant effect on cheese hardness when an extended denaturation of whey proteins took place under batch heating at 90 °C for 5 min.

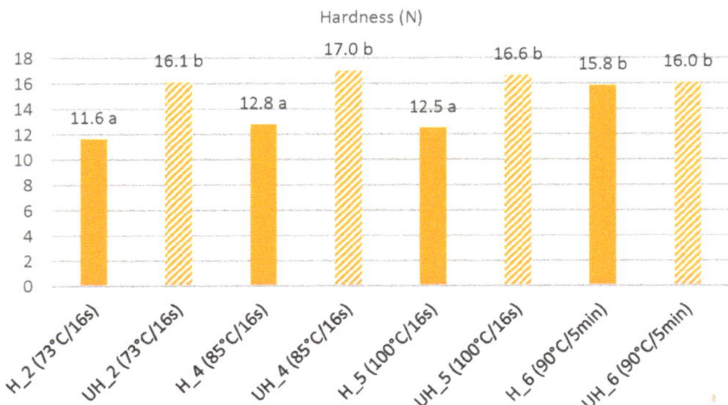

Figure 2. Interaction of homogenization and heating conditions on the hardness of Quark-type cheeses; a,b indicate significant differences (LSD, $p < 0.05$).

3.5. Proteolysis Indices of Cheese

The nitrogen (N) contents of the water-soluble fraction (WSN) and of the fraction soluble in 12% trichloroactic acid (TCAN) of cheeses are presented in Table 5. Storage at low temperature increased significantly ($p < 0.05$) both the WSN and TCAN. On average, WSN expressed as a percentage of total nitrogen (TN) increased from 10.07 ± 1.8 to $11.81 \pm 2.35\%$ from 5 to 20 d. Similarly, TCAN/TN increased from 4.49 ± 1.53 to $5.66 \pm 1.37\%$, indicating that secondary proteolysis took place at 4 °C, accumulating small/medium-size peptides and free amino acids [45]. All experimental factors and the homogenization–fat content and homogenization–heating interactions exhibited statistically significant effects ($p < 0.05$) on the proteolysis indices; therefore, they are grouped in Table 5 according to fat content and days of storage.

Table 5. Means of proteolysis indices of Quark-type cheeses made from unhomogenized and homogenized full- and reduced-fat milk heated under different conditions.

Index	%WSN		%WSN/TN		%TCAN		%TCAN/WSN	
Days	5	20	5	20	5	20	5	20
Full-Fat Cheese (FF)								
Homogenization								
H	0.166 [a]	0.194 [a]	10.29 [b]	13.11 [b]	0.061 [a]	0.086 [a]	37.52 [a]	44.04 [a]
UH	0.180 [b]	0.208 [b]	9.79 [a]	11.54 [a]	0.086 [b]	0.103 [b]	48.05 [b]	49.61 [b]
SE	0.0026	0.0045	0.161	0.408	0.0021	0.0025	1.28	1.05
LSD	0.0079	0.0136	0.482	1.22	0.0062	0.0074	3.84	3.15
Heat treatment								
2 (73 °C/16 s)	0.199 [d]	0.224 [c]	11.71 [d]	13.484 [c]	0.087 [c]	0.109 [c]	43.54 [c]	48.93 [b]
4 (85 °C/16 s)	0.182 [c]	0.209 [b,c]	10.80 [c]	13.147 [b,c]	0.085 [c]	0.105 [c]	49.59 [b]	50.76 [b]
5 (100 °C/16 s)	0.162 [b]	0.192 [a,b]	9.33 [b]	11.671 [a,b]	0.068 [b]	0.090 [b]	42.01 [b]	46.87 [b]
6 (90 °C/5 min)	0.148 [a]	0.180 [a]	8.32 [a]	10.997 [a]	0.054 [a]	0.074 [a]	36.01 [a]	40.73 [a]
SE	0.0037	0.0064	0.228	0.5781	0.0028	0.0035	1.81	1.48
LSD	0.0112	0.0191	0.683	1.733	0.0090	0.0105	5.42	4.45
Reduced-Fat Cheese (RF)								
Homogenization								
H	0.173 [a]	0.188 [a]	9.52 [a]	9.79 [a]	0.097 [b]	0.115 [b]	53.91 [b]	64.25 [b]
UH	0.225 [b]	0.252 [b]	10.91 [b]	12.89 [b]	0.077 [a]	0.092 [a]	36.20 [a]	38.60 [a]
SE	0.0090	0.0037	0.423	0.227	0.0055	0.0032	2.18	1.54
LSD	0.0251	0.0110	1.272	0.664	0.0167	0.0098	6.69	4.67

Table 5. *Cont.*

Index	%WSN		%WSN/TN		%TCAN		%TCAN/WSN	
Days	5	20	5	20	5	20	5	20
Heat treatment								
2 (73 °C/16 s)	0.219 [b]	0.253 [b]	11.21 [b]	13.04 [a]	0.095 [b]	0.106 [b]	36.96 [a]	42.03 [a]
4 (85 °C/16 s)	0.204 [b]	0.212 [b]	10.38 [b]	10.96 [c]	0.098 [b]	0.111 [b]	49.08 [b,c]	53.54 [b]
5 (100 °C/16 s)	0.212 [b]	0.229 [b]	11.12 [b]	12.08 [b]	0.069 [a]	0.088 [a]	40.06 [a,b]	52.25 [b]
6 (90 °C/5 min)	0.150 [a]	0.189 [a]	8.16 [a]	9.27 [d]	0.085 [a,b]	0.108 [b]	54.11 [c]	57.88 [b]
SE	0.0133	0.0052	0.620	0.302	0.0080	0.0043	3.11	2.17
LSD	0.0376	0.0155	1.879	0.966	0.0240	0.0139	9.84	6.61

Different letters indicate statistically significant differences (LSD, $p < 0.05$) in the group of means assigned to each experimental factor. H and UH, homogenized and unhomogenized cheese milk; WSN, water-soluble nitrogen; TCAN, nitrogen soluble in 12% trichloroacetic acid; SE, standard error; LSD, least significance difference.

The increase of the severity of heat treatment significantly ($p < 0.05$) decreased the WSN and WSN/TN ratio in accordance to the level of denaturation of major whey proteins in cheese milk (Table 1). Reduced proteolysis due to the increase of heating of Quark cheese milk has been also reported by Mara and Kelly [46]. Major whey proteins α-la and β-lg are the main WSN constituents of high moisture cheeses. As discussed earlier—in Sections 1, 3.2 and 3.3—heat-denatured whey proteins are complexed with casein micelles and MFGM, thus resulting in reduced nitrogen content of the soluble fraction. Moreover, the significant reduction of WSN observed in both FF and RF cheeses from homogenized milk (H) can be related to complexation with the caseins onto the surface of the homogenized MFGs. Secondary proteolysis in FF and RF cheeses corresponding to the TCAN fraction was not affected in a similar manner by the experimental factors (Table 5). In FF cheeses, the changes of TCAN paralleled those of WSN, decreasing by the increase of heat treatment and by the homogenization of cheese milk. Schematically, secondary proteolysis products included in the TCAN fraction are for the most part attributed to the further degradation of the products of primary proteolysis by the bacterial peptidases. Rennet is expected to be the major factor for primary proteolysis in the cheeses of the present study due to high moisture and low pH [45,46]. However, the low quantity of rennet added in cheese milk should be considered. The access of chymosin to caseins is hindered when they are complexed with denatured whey proteins. The casein matrix of homogenized milks contains also the new MFGs covered by caseins, and consequently, the paracasein matrix includes the homogenized MFGs e.g., [8,9,17]. These effects could be related to the lower proteolytic activity of chymosin post-manufacture, which reduces the primary proteolysis products used as substrates for the secondary changes. However, homogenization significantly ($p < 0.05$) increased the secondary proteolysis of RF cheeses. Since the increase of TCAN is mostly attributed to intracellular bacterial peptidases, a possible effect of RF cheese matrix on the autolysis of the bacterial cells can be taken into account. Similar to our results for RF cheese are the findings of Zamora et al. [47] for a starter-free cheese model with approximately pH 6.7 and moisture 68%, who report that the homogenization of milk at 18–2 MPa decreased WSN/TN but increased free amino groups. The same group [48] studied the effect of homogenization at 18–2 MPa of cheese milk on major proteolytic factors in soft rennet–curd goat milk cheese with 4.6–4.9. They reported that residual chymosin was not affected by homogenization, while the aminopeptidase activity and respective WSN/TN and free amino groups were significantly lower ($p < 0.05$) at the start of ripening.

3.6. Organoleptic Evaluation of Cheese

The results of the organoleptic evaluation of cheeses at 20 d of storage are shown in Table 6. The experimental factors and their interactions did not statistically significantly affect the scores with the exception of the significant effect ($p < 0.05$) of fat content on the appearance of cheeses. On storage, no expulsion of whey from cheese mass was observed. Flavor and textural defects were not reported.

Table 6. Means of organoleptic scores of Quark-type cheeses at 20 d of storage made from unhomogenized and homogenized full- and reduced-fat milk heated under different conditions. Different letters indicate statistically significant differences (LSD, $p < 0.05$) in the group of means assigned to each experimental factor. H and UH, homogenized and unhomogenized cheese milk; FF, full fat; RF, reduced fat; SE, standard error; LSD, least significance difference.

Factors	Appearance 0–10	Texture 0–40	Flavour 0–50	Total Score 0–100
Homogenization				
H	9.04	32.73	39.47	77.94
UH	8.98	33.08	38.96	78.37
SE	0.07	0.50	1.06	2.41
LSD	0.21	1.45	3.02	6.90
Fat-content				
FF	8.84 [a]	32.78	38.50	79.60
RF	9.18 [b]	33.04	39.93	76.71
SE	0.07	0.50	1.06	2.41
LSD	0.21	1.45	3.02	6.9
Heat treatment				
2 (73 °C/16 s)	9.00	33.03	37.14	75.23
4 (85 °C/16 s)	8.95	32.56	40.00	79.87
5 (100 °C/5 min)	9.02	33.16	39.07	78.31
6 (90 °C/16 s)	9.06	32.90	40.65	79.22
SE	0.10	0.70	1.46	3.34
LSD	0.29	2.01	4.27	9.50

The lack of significant differences indicates that the interventions applied in the Quark-type cheese making in the frame of the present study could be acceptable in organoleptic terms despite their above-mentioned profound effects on cheese biochemical and textural characteristics.

4. Conclusions

Heat treatments and the homogenization of cheese milk affected significantly ($p < 0.05$) the composition and texture of the experimental Quark-type cheeses. The increase of the intensity of heat treatment increased moisture and protein retention in the curd, and the same was true for homogenization; the effect of the latter was more pronounced in full-fat cheeses. However, the effects were limited when cheese milk was treated under conditions that induced a high level of α-lactalbumin denaturation. Moreover, under these conditions, significantly ($p < 0.05$) higher residual lactose and citric acid contents were observed, implying a retardation of starter activity. Storage at 4 °C did not affect composition and texture but increased proteolysis.

Therefore, heat treatment conditions of cheese milk that induce a considerable denaturation of β-lactoglobulin and leave a considerable amount of native α-lactalbumin were more adequate for the manufacture of this cheese type, whereas homogenization was more effective for full-fat cheeses.

In conclusion, a "clean label", paste-like, full- or reduced-fat, high-moisture cheese from combined acid–rennet curd can be manufactured by combining physical treatments of cow cheese milk.

Author Contributions: Conceptualization, G.M.; Data curation, S.L., E.Z. and G.M.; Formal analysis, S.L., E.Z. and D.L.; Investigation, S.L., E.Z. and D.L.; Methodology, E.Z., E.M. and G.M.; Project administration, G.M.; Resources, E.M. and G.M.; Software, G.M.; Supervision, G.M.; Validation, G.M.

and G.M.; Visualization, E.Z. and G.M.; Writing—original draft, S.L., E.Z. and G.M.; Writing—review and editing, G.M. All authors have read and agreed to the published version of the manuscript.

Funding: This research received no external funding.

Data Availability Statement: The data presented in this study are available on request from the corresponding author.

Acknowledgments: The authors sincerely thank Theodoros Paschos and Lambros Sakkas for their contribution to the experimental milk treatments. A great part of the present study has been carried out in the framework of the Programme of Postgraduate Studies entitled 'Integrated Production Management of Milk and Dairy Products'.

Conflicts of Interest: The authors declare no conflict of interest.

References

1. Schulz-Collins, D.; Senge, B. *Acid- and Acid/Rennet-Curd Cheeses Part A: Quark, Cream Cheese and Related Varieties Cheese: Chemistry, Physics and Microbiology*; Fox, P.F., McSweeney, P.L.H., Cogan, T.M., Guinee, T.P., Eds.; Academic Press: London, UK, 2004; Volume 2, pp. 301–328.
2. Farkye, N.Y. Quark, Quark-like Products, and Concentrated Yogurts. In *Cheese, Chemistry, Physics & Microbiology*, 4th ed.; McSweeney, P.L.H., Fox, P.F., Cotter, P.D., Everett, D.V., Eds.; Academic Press is an Imprint of Elsevier: London, UK, 2017; pp. 1103–1110.
3. Codex Alimenatrius. *Milk and Milk Products*, 2nd ed.; World Health Organization Food and Agriculture Organization of the United Nations: Rome, Italy, 2011.
4. Guinee, T.P.; McSweeney, P.L.H. Significance of Milk Fat in Cheese. In *Advanced Dairy Chemistry*, 3rd ed.; Fox, P.F., McSweeney, P.L.H., Eds.; Springer: New York, NY, USA, 2006; Volume 2, pp. 377–440.
5. Farkye, N.Y.; Guinee, T.P. Low-Fat and Low-Sodium Cheeses. In *Cheese, Chemistry, Physics & Microbiology*, 4th ed.; McSweeney, P.L.H., Fox, P.F., Cotter, P.D., Everett, D.V., Eds.; Academic Press is an imprint of Elsevier: London, UK, 2017; pp. 699–714.
6. Sheth, H.; Jelen, P.; Ozimek, L.; Sauer, W. Yield, sensory properties, and nutritive qualities of quarg produced from lactose-hydrolyzed and high heated milk. *J. Dairy Sci.* **1988**, *71*, 2891–2897. [CrossRef]
7. Kelly, A.L.; O'Donnell, H.J. Composition, gel properties and microstructure of quarg as affected by processing parameters and milk quality. *Intern. Dairy J.* **1998**, *8*, 295–301. [CrossRef]
8. Guyomarc'h, F. Formation of heat-induced protein aggregates in milk as a means to recover the whey protein fraction in cheese manufacture, and potential of heat-treating milk at alkaline pH values in order to keep its rennet coagulation properties. *A Rev. Lait* **2006**, *86*. [CrossRef]
9. Kelly, A.L.; Huppertz, T.; Sheehan, J.J. Pre-treatment of cheese milk: Principles and developments. *Dairy Sci. Technol.* **2008**, *88*, 549–572. [CrossRef]
10. Moatsou, G.; Zoidou, E.; Choundala, E.; Koutsaris, K.; Kopsia, O.; Thergiaki, K.; Sakkas, L. Development of reduced-fat, reduced-sodium semi-hard sheep milk cheese. *Foods* **2019**, *8*, 204. [CrossRef] [PubMed]
11. Lucey, J.A.; Tamehana, M.; Singh, H.; Munro, P.A. Effect of heat treatment on the physical properties of milk gels made with both rennet and acid. *Intern. Dairy J.* **2001**, *11*, 559–565. [CrossRef]
12. Vaziri, M.; Abbasi, H.; Mortazavi, A. Microstructure and physical properties of quarg cheese as affected by different heat treatments. *J. Food Process. Preserv.* **2010**, *34*, 2–14. [CrossRef]
13. Frau, F.; Font de Valdez, G.; Pece, N. Effect of pasteurization temperature, starter culture, and incubation temperature on the physicochemical properties, yield, rheology, and sensory characteristics of spreadable goat cheese. *J. Food Process.* **2014**, 1–8. [CrossRef]
14. Miloradovic, Z.; Miocinovic, J.; Kljajevic, N.; Tomasevic, I.; Pudja, P. The influence of milk heat treatment on composition, texture, colour and sensory characteristics of cows' and goats' Quark-type cheeses. *Small Rumin. Res.* **2018**, *169*, 154–159. [CrossRef]
15. Ong, L.; Dagastine, R.R.; Kentish, S.E.; Gras, S.L. The effect of milk processing on the microstructure of the milk fat globule and rennet induced gel observed using confocal laser scanning microscopy. *J. Food Sci.* **2010**, *75*, E135–E145. [CrossRef]
16. Ong, L.; Dagastine, R.R.; Kentish, S.E.; Gras, S.L. Microstructure of milk gel and cheese curd observed using cryo-scanning electron microscopy and confocal microscopy. *LWT-Food Sci. Technol.* **2011**, *44*, 1291–1302. [CrossRef]
17. Obeid, A.; Guyomarch, F.; Tanguy, G.; Leconte, N.; Rousseau, F.; Dolivet, A.; Leduc, A.; Wu, X.; Cauty, C.; Jan, G.; et al. The adhesion of homogenized fat globules to proteins is increased by milk heat treatment and acidic pH: Quantitative insights provided by AFM force spectroscopy. *Food Res. Intern.* **2020**, *129*, 108847. [CrossRef] [PubMed]
18. Obeid, S.; Guyomarc'h, F.; Francius, G.; Guillemin, H.; Wu, X.; Pezennec, S.; Famelart, M.-H.; Cauty, C.; Gaucheron, F.; Lopez, C. The surface properties of milk fat globules govern their interactions with the caseins: Role of homogenization and pH probed by AFM force spectroscopy. *Colloids Surf. B Biointerfaces* **2019**, *182*, 110363. [CrossRef] [PubMed]
19. Ningtyas, D.W.; Bhandari, B.; Bansal, N.; Prakash, S. Effect of homogenization of cheese milk and high-shear mixing of the curd during Cream cheese manufacture. *Intern. J. Dairy Technol.* **2018**, *71*, 417–431. [CrossRef]
20. Kiokias, S.; Bot, A. Effect of protein denaturation on temperature cycling stability of heat-treated acidified protein-stabilised o/w emulsions. *Food Hydrocoll.* **2005**, *19*, 493–501. [CrossRef]

21. Brighenti, M.; Govindasamy-Lucey, S.; Jaeggi, J.J.; Johnson, M.E.; Lucey, J.A. Effects of processing conditions on the texture and rheological properties of model acid gels and Cream cheese. *J. Dairy Sci.* **2018**, *101*, 6762–6775. [CrossRef] [PubMed]

22. Coutouly, A.; Riaublanc, A.; Axelos, M.; Gaucher, I. Effect of heat treatment, final pH of acidification, and homogenization pressure on the texture properties of Cream cheese. *Dairy Sci. Technol.* **2014**, *94*, 125–144. [CrossRef]

23. Brighenti, M.; Govindasamy-Lucey, S.; Lim, K.; Nelson, K.; Lucey, J.A. Characterization of the rheological, textural, and sensory properties of samples of commercial US Cream cheese with different fat contents. *J. Dairy Sci.* **2008**, *91*, 4501–4517. [CrossRef]

24. Nguyen, P.T.M.; Bhandari, B.; Prakash, S. Tribological method to measure lubricating properties of dairy products. *J. Food Eng.* **2017**, *168*, 27–34. [CrossRef]

25. Macdougall, P.E.; Ong, L.; Palmer, M.W.; Gras, S.L. The microstructure and textural properties of Australian Cream cheese with differing composition. *Intern. Dairy J.* **2019**, 104548. [CrossRef]

26. Moschopoulou, E.; Moatsou, G. Greek Dairy Products: Composition and Processing. In *Mediterranean Food: Composition & Processing*; da Cruz, R.M.S., Vieira, M.M.C., Eds.; CRC Press, Taylor & Francis Group: Boca Raton, FL, USA; pp. 268–320.

27. Katsiari, M.; Kondyli, E.; Voutsinas, L. The quality of Galotyri-type cheese made with different starter cultures. *Food Control.* **2009**, *20*, 113–118. [CrossRef]

28. Kondyli, E.; Katsiari, M.; Voutsinas, L.P. Chemical and sensory characteristics of Galotyri-type cheese made using different producers. *Food Control.* **2008**, *19*, 301–307. [CrossRef]

29. Kondyli, E.; Massouras, T.; Katsiari, M.C.; Voutsinas, L.P. Lipolysis and volatile compounds of Galotyri-type cheese made using different procedures. *Small Rumin. Res.* **2013**, *113*, 432–436. [CrossRef]

30. Zoidou, E.; Karageorgos, D.; Massouras, T.; Anifantakis, E. The effect of probiotic lactic acid bacteria on the characteristics of Galotyri cheese. *Intern. J. Clin. Nutr. Diet.* **2016**, *2*, 114. [CrossRef]

31. Sakkas, L.; Moutafi, A.; Moschopoulou, E.; Moatsou, G. Assessment of heat treatment of various types of milk. *Food Chem.* **2014**, *159*, 293–301. [CrossRef] [PubMed]

32. Kaminarides, S.; Stamou, P.; Massouras, T. Changes of organic acids, volatile aroma compounds and sensory characteristics of Halloumi cheese kept in brine. *Food Chem.* **2007**, *100*, 219–225. [CrossRef]

33. International Standard ISO22662/IDF198. *Milk and Milk Products-Determination of Lactose Content by High Performance Liquid Chromatography (Reference Method)*; International Dairy Federation: Brussels, Belgium, 2007.

34. Kaminarides, S.; Anifantakis, E. Characteristics of set type yoghurt made from caprine or ovine milk and mixtures of the two. *Intern. J. Food Sci. Technol.* **2004**, *39*, 319–324.

35. Gunasekaran, S.; Ak, M.M. Cheese Texture. In *Cheese Rheology and Texture*, 1st ed.; CRC Press LLC: Boca Raton, FL, USA, 2003; pp. 308–338.

36. Moschopoulou, E.; Sakkas, L.; Zoidou, E.; Theodorou, G.; Sgouridou, E.; Kalathaki, C.; Liarakou, A.; Chatzigeorgiou, A.; Politis, I.; Moatsou, G. Effect of milk kind and storage on the biochemical textural and biofunctional characteristics of set-style yoghurt. *Intern. Dairy J.* **2018**, *77*, 47–55. [CrossRef]

37. Walstra, P.; Wouters, J.T.M.; Geurts, T.J. Changes in Salts. In *Dairy Science and Technology*, 2nd ed.; CRC Press Taylor and Francis Group: Boca Raton, FL, USA, 2006; pp. 54–58.

38. McSweeney, P.L.H.; Fox, P.F.; Coccia, F. Metabolism of Residual Lactose and of Lactate and Citrate. In *Cheese, Chemistry, Physics & Microbiology*, 4th ed.; McSweeney, P.L.H., Fox, P.F., Cotter, P.D., Everett, D.V., Eds.; Academic Press is an Imprint of Elsevier: London, UK, 2017; pp. 411–421.

39. Zeppa, G.; Rolle, L. A study on organic acid, sugar and ketone contents in typical piedmont cheeses. *Ital. J. Food Sci.* **2008**, *20*, 127–139.

40. Papadakis, E.N.; Polychroniadou, A. Application of a microwave-assisted extraction method for the extraction of organic acids from Greek cheeses and sheep milk yoghurt and subsequent analysis by ion-exclusion liquid chromatography. *Intern. Dairy J.* **2005**, *15*, 65–172. [CrossRef]

41. Ozturkoglu-Budak, S.; Akal, H.C.; Türkmen, N. Use of kefir and buttermilk to produce an innovative quark cheese. *J. Food Sci. Technol.* **2020**. [CrossRef]

42. Skryplonek, K.; Dmytrów, I.; Mituniewicz-Małek, A. The use of buttermilk as a raw material for cheese production. *Int. J. Dairy Technol.* **2019**, *72*, 610–616. [CrossRef]

43. Solowiej, B. Textural, rheological and melting properties of acid casein reduced-fat processed cheese analogues. *Milchwiss.* **2012**, *67*, 9–13.

44. Kealy, T. Application of liquid and solid rheological technologies to the textural characterisation of semi-solid foods. *Food Res. Intern.* **2006**, *39*, 265–276. [CrossRef]

45. Nega, A.; Moatsou, G. Proteolysis and related enzymatic activities in ten Greek cheese varieties. *Dairy Sci. Technol.* **2012**, *92*, 57–73. [CrossRef]

46. Mara, O.; Kelly, A.L. Contribution of milk enzymes, starter and rennet to proteolysis during storage of quarg. *Intern. Dairy J.* **1998**, *8*, 973–979. [CrossRef]

47. Zamora, A.; Juan, B.; Trujillo, A.J. Compositional and biochemical changes during cold storage of starter-free fresh cheeses made from ultra-high-pressure homogenised milk. *Food Chem.* **2015**, *176*, 433–440. [CrossRef] [PubMed]

48. Juan, B.; Zamora, A.; Quevedo, J.M.; Trujillo, A.-J. Proteolysis of cheese made from goat milk treated by ultra high pressure homogenization. *LWT-Food Sci. Technol.* **2016**, *69*, 17–23. [CrossRef]

Article

Molecular Properties of *Flammulina velutipes* Polysaccharide–Whey Protein Isolate (WPI) Complexes via Noncovalent Interactions

Jiaqi Shang [1,2], Minhe Liao [1,2], Ritian Jin [1,2], Xiangyu Teng [1,2], Hao Li [1,2], Yan Xu [1,2], Ligang Zhang [1,2,*] and Ning Liu [1,2,*]

[1] Key Laboratory of Dairy Science, Ministry of Education, Harbin 150030, China; ashang10@outlook.com (J.S.); minheliao@outlook.com (M.L.); ritianjin@outlook.com (R.J.); txy957180709@gmail.com (X.T.); bluesboylee@outlook.com (H.L.); yanxu1991@neau.edu.cn (Y.X.)

[2] College of Food Science, Northeast Agricultural University, Harbin 150030, China

* Correspondence: zhang@neau.edu.cn (L.Z.); liuning@neau.edu.cn (N.L.); Tel.: +86-451-55191827 (L.Z. & N.L.)

Abstract: Whey protein isolate (WPI) has a variety of nutritional benefits. The stability of WPI beverages has attracted a large amount of attention. In this study, *Flammulina velutipes* polysaccharides (FVPs) interacted with WPI to improve the stability via noncovalent interactions. Multiple light scattering studies showed that FVPs can improve the stability of WPI solutions, with results of radical scavenging activity assays demonstrating that the solutions of the complex had antioxidant activity. The addition of FVPs significantly altered the secondary structures of WPI, including its α-helix and random coil. The results of bio-layer interferometry (BLI) analysis indicated that FVPs interacted with the WPI, and the equilibrium dissociation constant (K_D) was calculated as 1.736×10^{-4} M in this study. The in vitro digestibility studies showed that the FVPs protected WPI from pepsin digestion, increasing the satiety. Therefore, FVPs effectively interact with WPI through noncovalent interactions and improve the stability of WPI, with this method expected to be used in protein-enriched and functional beverages.

Keywords: *Flammulina velutipes*; protein–polysaccharide complexes; stability; bio-layer interferometry; in vitro digestibility; binding regions

1. Introduction

Whey protein is a natural byproduct of cheese production and has high nutritional value. The molecular structure of whey protein has abundant branched-chain amino acids (BCAA; leucine, isoleucine, and valine), and its total amino acid composition is similar to that of skeletal muscles [1–3]. Whey protein isolate (WPI) is a class of whey protein that is often added to fitness, exercise nutrition, and weight management foods [4,5]. Many WPI-containing beverages are subjected to low pH (pH < 3.5), and the astringency is obvious at low pH [6]. The pH range with the best taste for WPI beverages at low pH is 4.0–6.0, as the isoelectric point (pI) of WPI is 5.1; however, making WPI stable near its pI is a problem.

It is now well recognized that the stability of proteins can be improved by conjunction with polysaccharides via covalent or noncovalent interactions [7–9]. Covalent interactions always occur through Maillard reactions, while noncovalent interactions result mainly from hydrophobic interactions, hydrogen binding, or electrostatic attraction under various conditions [10–12]. Maillard reactions in food may lead to the nutrients loss, and the contents of some proteins in these food are decreased or become non-digestible [13]. High temperature, time-consuming nature, and other parameters in the generation of food products may have harmful effects to health, such as mutagenicity and metabolic diseases [14]. When the pH of the reaction system is lower than the pI of protein, a strong electrostatic attraction occurs between the inversely charged protein and polysaccharide,

Citation: Shang, J.; Liao, M.; Jin, R.; Teng, X.; Li, H.; Xu, Y.; Zhang, L.; Liu, N. Molecular Properties of *Flammulina velutipes* Polysaccharide–Whey Protein Isolate (WPI) Complexes via Noncovalent Interactions. *Foods* **2021**, *10*, 1. https://dx.doi.org/10.3390/foods10010001

Received: 20 October 2020
Accepted: 18 December 2020
Published: 22 December 2020

Publisher's Note: MDPI stays neutral with regard to jurisdictional claims in published maps and institutional affiliations.

Copyright: © 2020 by the authors. Licensee MDPI, Basel, Switzerland. This article is an open access article distributed under the terms and conditions of the Creative Commons Attribution (CC BY) license (https://creativecommons.org/licenses/by/4.0/).

thus forms a strong electrostatic complex (the pI is the pH value at which the surface of a molecule is uncharged) [15]. Therefore, the electrostatic attraction can make the protein and polysaccharide complexes keep the nutritional values without promoting detrimental health effects.

Previous studies have studied the interactions between polysaccharides and proteins. It has been reported that the presence of dextran sulfate (DS) decreased the turbidity of heated β-lactoglobulin (β-LG) and improved heat stability of β-LG [8]. However, changes in noncovalent interactions on structures of proteins in the composite system have yet to be resolved, and few studies have investigated the noncovalent interactions in the complex of WPI and bioactive polysaccharides. *Flammulina velutipes* belongs to Tricholomataceae and is widely eaten in China and Japan [16]. *Flammulina velutipes* polysaccharides (FVPs) are regarded as one of the most important substances in *Flammulina velutipes*, which have been reported to possess a wide spectrum of biological activities including antioxidant and hepatoprotective activities, the ability to prevent Hepatitis B Virus (HBV) infection, and the capacity to induce immunity by activating macrophages [17–20]. The molecular structure of polysaccharides is usually composed of many hydrophilic groups [21]. Hence, polysaccharides could have a beneficial impact on the stability of the WPI system.

The aim of this study was to investigate whether FVP could improve stability of WPI by occurring noncovalent interaction. In this study, FVPs were used as a stabilizer of the WPI system. The influence of adding FVP on particles properties and stability of WPI in aqueous solution was investigated. In addition, we investigated the antioxidant activity via the scavenging activity of three radicals. The changes of structure in WPI after noncovalent interaction with FVPs occurred were indicated. Then, we verified that there was an interaction between FVPs and WPI via the reaction kinetics. Finally, the digestibility and binding sites were assessed to understand the interaction between FVPs and WPI more stereoscopically. The results of this study should facilitate the development of new methods of WPI stabilizing with applications in the food industry.

2. Materials and Methods

2.1. Materials

The *Flammulina velutipes* used in this study originated from the Chinese mainland and were purchased from a B.U.T. Mart in Harbin. WPI was from Davisco Foods International (Le Sueur, MN, USA). Nitro blue tetrazolium (NBT) and phenazine methosulfate (PMS) were from Sigma-Aldrich (St. Louis, MO, USA). EZ-Link NHS-PEG$_{12}$-Biotin (21312) and trypsin (MS-grade, 90057, $\geq$15,000 u/mg) were from Thermo Scientific (Waltham, MA, USA). Other chemicals were purchased from the Aladdin reagent official website.

2.2. Extraction of FVPs

The FVPs extraction was performed according to a previously described method with vacuum concentration [22,23]. In brief, *Flammulina velutipes* powder was extracted with distilled water at 85 °C for 2 h. Then, the extracts were concentrated at 50 °C. After concentration, the extracts were centrifuged at 4000 rpm for 30 min. Subsequently, the supernatant was collected and precipitated using 4 volumes of absolute ethanol. The resulting precipitate was collected and freeze-dried (crude polysaccharide). Proteins were removed from the sediment using the Sevag method. Briefly, Sevag reagent contained chloroform and n-butanol solvents at a volume ratio of 1:5. A total of 3 mg/mL crude polysaccharide solution was added with the Sevag reagent at a volume ratio of 1:3. The well-mixed solution was stirred for 20 min and stood undisturbed to form the phase-separation. The supernatant was collected, and 3 steps were repeated 5 times. Furthermore, the samples were decolorized with activated carbon after deproteinization. Finally, the FVPs were filtered by air pump filtration and dried on a freeze dryer (Marin Christ Alpha 2–4 LSCplus, Osterode, Germany).

2.3. Preparation of FVP–WPI Complex Solutions

The FVP–WPI complex solutions were prepared according a previously described method with different ratios [24]. The solutions of FVP–WPI complex with final concentrations of 0.4% (*w/w*) FVPs and 3–7% (*w/w*) WPI were obtained using a 0.8% FVP stock solution and 6–14% WPI stock solutions, which were stored at 4 °C overnight. FVP and WPI stock solutions were mixed at the weight ratio of 1:1. The pH of solutions was adjusted to 4.5 with 1 M NaOH and 1 M HCl, after which 20 mL of each solution was transferred to a flat-bottomed cylindrical glass measurement cell with a black cap for stability analysis.

2.4. Particle Size and Zeta Potential

The apparent size and zeta potential of the samples were analyzed by Dynamic Light Scattering (Malvern Zetasizer Nano ZS, Worcestershire, UK) using a previously described method [25]. Distilled water was used to dilute the sample solutions at a ratio of 1:200, and the pH of solutions was adjusted to 4.5 with 1 M NaOH and 1 M HCl. Samples were well mixed before measurement. Each sample was measured 3 times, and the results were averaged. The cell was maintained at room temperature.

2.5. Stability Monitoring

Stability was monitored using multiple light scattering (Formulaction TURBISCAN TOWER, Toulouse, France) by using a described scheme of the equipment [26]. Multiple light scattering is used to comprehensively characterize the stability characteristics of high-concentration dispersion systems. The samples were scanned from the bottom to the top in order to monitor the optical properties of the dispersion along the height in the cell. The sample was stored at 4 °C and monitored after 0, 5, 10, and 15 d. Each monitoring duration was 30 min, and scanning was performed once in every 110 s for 17 times totally. The stability of the samples was deduced from the back scattering (BS) data.

2.6. Antioxidant Activity

2.6.1. ABTS Radical Scavenging Activity

The 2,2′-azinobis (3-ethylbenzothiazoline-6-sulphonic acid) (ABTS) radical scavenging activities of the samples were determined according to a previously described method with 7 mM ABTS [27]. In brief, an adequate volume of 7 mM ABTS was added to the same volume of 2.45 mM potassium persulfate solution, after which the mixture was incubated in the dark for 12 h at room temperature before being diluted with PBS (phosphate-buffered saline, 0.01 M, pH 7.4) to attain an absorbance of the working solution of 0.70 ± 0.02 at 734 nm. Then, 1.0 mL sample was mixed with 4.0 mL of the ABTS working solution and vortexed. The absorbance of the final mixture was read at 734 nm using the microplate reader (Molecular Devices M2e, San Francisco, CA, USA), and the ABTS radical scavenging activity was calculated using Equation (1).

$$\text{ABTS radical scavenging activity (\%)} = [1 - (A_1 - A_2)/A_0] \times 100 \tag{1}$$

where A_0 is the Abs_{734} of the mixture without sample, A_1 is the Abs_{734} of the sample, and A_2 is the Abs_{734} of the sample without ABTS radical solution.

2.6.2. Hydroxyl Radical Scavenging Activity

The hydroxyl radical scavenging activities of the samples were determined by a reported method [28]. In brief, 0.5 mL of 9 mM ferrous sulfate solution was mixed with 1.0 mL of 8.8 mM hydrogen peroxide solution to perform the Fenton reaction. Then, 1 mL sample solution was added to the Fenton reaction mixture. Finally, 0.1 mL 9 mM salicylic acid (dissolved in ethanol) was added to the mixture and vortexed, after which the mixture was incubated at 37 °C for 1 h. The absorbance of the final mixture was read at

510 nm on a microplate reader, and the hydroxyl radical scavenging activity was calculated using Equation (2).

$$\text{Hydroxyl radical scavenging activity (\%)} = [1 - (A_1 - A_2)/A_0] \times 100 \qquad (2)$$

where A_0 is the Abs_{510} of the control (without sample), A_1 is the Abs_{510} of the sample with salicylic acid, and A_2 is the Abs_{510} of the sample without salicylic acid.

2.6.3. Superoxide Anion Radical Scavenging Activity

The superoxide anion radical scavenging activities of the samples were determined using a reported method [29]. Briefly, 1 mL sample solution was mixed with 1 mL 300 μM NBT, after which 1 mL 936 μM NADH (nicotinamide adenine dinucleotide) was added to the mixture. Finally, 1 mL 120 μM PMS was added to the reaction mixture before it was incubated at 25 °C for 5 min. The absorbance of the final mixture was read at 560 nm, and the superoxide anion radical scavenging activity was calculated using Equation (3).

$$\text{Superoxide anion scavenging activity (\%)} = (A_0 - A_1)/A_0 \times 100 \qquad (3)$$

where A_0 is the Abs_{560} of the control (without sample), and A_1 is the Abs_{560} of the sample.

2.7. FTIR

FTIR analysis was performed using an FTIR spectrometer (NICOLET, Madison, USA) according to the method reported by Yan et al. [29]. The lyophilized powders of FVPs, WPI, and FVP–WPI were grounded with KBr powder and then pressed into pellets for FTIR measurements. FTIR spectra were recorded at a wavelength range of 400–4000 cm^{-1}.

2.8. Raman Spectroscopy

Raman spectra of FVPs, WPI, and 0.4% FVP–5% WPI were conducted using a Raman spectrometer (HORIBA Evilution, France) equipped with a microscope and a 532 nm near-infrared diode laser according to a reported method with different detection ranges [30]. Each sample was deposited onto a microscope slide and freeze-dried prior to Raman measurement. The Raman spectra were recorded at room temperature with a detection range from 400–2800 cm^{-1}. Spectral data were collected using LabSpec6 and were baseline corrected and normalized according to the protein phenylalanine peak at 1003 ± 1 cm^{-1}.

2.9. XRD

The XRD spectra were performed according to Liu et al., with different ranges of scan [31]. The XRD spectra were performed using an X-ray diffractometer (Fangyuan 2700, Dandong, China) at 40 mA and 40 kV, where the target material was Cu. The scattering angle (2θ) was scanned over the range of 5–80° at a scanning step of 0.05°. Spectral data were collected and further analyzed using MDJ Jade 6.0.

2.10. Reaction Kinetics

Recently, bio-layer interferometry (BLI) was developed to assess molecular interactions [32,33]. The reaction kinetics were determined using an unlabeled molecular interaction analysis system (Pall ForteBio Octet RED96e, Fremont, CA, USA). The biotinylation of WPI was performed using EZ-Link NHS-PEG$_{12}$-Biotin according to the specification, where 1 mL 1 mg/mL WPI stock solution was mixed with 3 μL of 10 mM EZ-Link NHS-PEG$_{12}$-Biotin mother liquor and vortexed. The reaction mixtures were incubated at room temperature for 30 min, after which the free biotin was removed using a Zeba desalination centrifugal column (Thermo Scientific). Then, 600 μL biotinylated WPI was added to a preprocessed desalination centrifugal column and centrifuged at 1000× g for 2 min to collect the target protein. The protein concentration was determined using a BCA Protein Assay kit according to the manufactures' instructions (Meilunbio MA0082, Dalian, China). The biotinylated WPI was diluted to 20 μg/mL in PBS (0.01 M). The test was conducted using a previously reported method without running buffer (containing BSA) [34,35]. In each assay, the streptavidin (SA) biosensor was balanced with PBS for 60 s before being loaded with biotinylated WPI for 120 s to achieve a loading signal between 2.0 and 2.1 nm, which was followed by a balancing step in PBST (0.01 M PBS + 0.02% Tween) for 120 s. Subsequently, association of WPI with FVPs at concentrations of 31.25, 62.5, 125, 250, 500, and 1000 µg/mL were performed for 240 s. Finally, the dissociation was monitored in PBST for 120 s.

2.11. Protein Digestion

In vitro pepsin digestion in this study was performed according to methods reported by Akl et al. [36]. Simulated gastric fluid (SGF) was composed of 0.2% NaCl, 0.32% pepsin ($\geq$250 u/mg), and 0.7% HCl (36.5%) according to the United States Pharmacopeial Convention. After diluting 7 mL of HCl in an appropriate volume of water, we added 3.2 g of pepsin and 2 g NaCl to the HCl solution. Then, the mixture was diluted to 1000 mL. SGF was prepared immediately before the digestion. First, 10 mL SGF was incubated with continuous shaking at 95 rpm for 5 min at 37 °C in a temperature-controlled incubator. Then, 10 mL WPI or 0.4% FVP–5% WPI sample solution was added into the reaction system. The pH of the solution was adjusted to 7.0 with 1 M NaOH after reacting for 10 or 120 min, respectively.

Trypsin digestion was carried out according to methods promoted by Zhang et al. with minor modifications [37]. First, the pepsin-digested samples were diluted to 10 µg/µL. Then, 20 µL diluted samples were mixed with 40 µL of protein denaturation solution (6 M GdmCl, 100 mM Tris, 10 mM Tris-(2-cyanoethyl) phophine, 40 mM Bis (2,4-pentanedionato) calcium, pH 8.5). After being heated at 95 °C for 5 min, the solution was transferred to an ultrafiltration tube (molecular weight cutoff: 10 kDa; Millipore) after cooling and centrifuged at 14,000 rpm for 15 min. The peptide fragments greater than 10 kDa were washed twice with 50 mM ammonium bicarbonate solution. Then, trypsin was added to the solution with a final ratio 50:1 (*w*/*w*, protein/enzyme), and the well-mixed solution was reacted at 37 °C for 16 h. Finally, the mixture in the ultrafiltration tube was centrifuged at 14,000 rpm for 15 min, and the effluents were used for proteomic analysis. Triple replicates were performed for each sample.

2.12. Proteomic Analysis

A nano-liquid phase system (Thermo Scientific Easy nLC1200, Waltham, MA, USA) was used to isolate the peptides according to the manufacturer's instructions. Solvent A was water (0.1% formic acid), and solvent B was 80% acetonitrile (0.1% formic acid). The chromatographic column was equilibrated using mobile phase A, and the samples were loaded by an automatic sampler and separated at a flow rate of 300 nL/min using gradient conditions described in Table 1. The samples were analyzed by mass spectrometry (Thermo Scientific Q-EXACTIVE PLUS, Waltham, MA, USA), with survey scans of peptide precursors performed from 350 to 1800 *m/z* at 200 *m/z*. After each survey scan, the AGC target for MS/MS was set to 3×10^6, and the maximum injection time was limited to 50 ms.

Table 1. The gradient conditions of the nano liquid phase system.

Composition	Time
0–8% B	0–5 min
8–32% B	5–75 min
32–90% B	75–77 min
90–0% B	82–85 min
0% B	85–90 min

All data were obtained in *. raw format, and the data files were analyzed against transforming growth factor beta-1 proprotein (P18341), transforming growth factor beta-2 proprotein (P21214), α-lactalbumin (P00711), β-lactoglobulin, (P02754), serum albumin

(P02769), lactoperoxidase (P80025), and lactotransferrin (P24627), downloaded from the UniProt database. After the analysis of the quantification data in MaxQuant to obtain the "Peptides.txt" file, intensities of the "Peptides.txt" data were $\log_2$ transformed. Then, a *t*-test was performed to analyze the significance using Perseus.

2.13. Statistical Analysis

Each experiment was performed 3 times. Results were presented as mean value ± standard deviation (SD). Statistical analysis of the results was performed by one-way analysis of variance (ANOVA) and Duncan's test using the SPSS software. Values of $p < 0.05$ were considered to be statistically significant.

3. Results

3.1. Characterization of WPI and FVP–WPI Particles

The formation of FVP–WPI complex changed the properties of WPI particles, which can be assessed by the measurement of the particle size and zeta potential. The particle size of WPI was 189.68 nm at pH 7.0, and when the pH was adjusted to 4.5, the particle size of WPI increased to 1586.45 ± 70.67 nm (3% WPI), 1970.33 ± 88.85 nm (4% WPI), 1336 ± 58.05 nm (5% WPI), 1709.33 ± 36.30 nm (6% WPI), and 1557.00 ± 63.27 nm (7% WPI). This phenomenon may be attributed to the pH near the pI of WPI. The apparent sizes of the FVP–WPI complexes and WPI at pH 4.5 are presented in Figure 1A, which showed that the particle sizes for the FVP–WPI complexes were smaller than those measured in solutions with WPI alone. Furthermore, smallest particle sizes were observed for WPI and FVP–WPI complexes obtained with a 5% WPI concentration, for which particle sizes in the solutions of WPI and solutions of FVP–WPI complex were approximately 1336 ± 58.05 and 707.13 ± 26.17 nm, respectively. Zhao et al. reported that the addition of soybean soluble polysaccharides and beet pectin to lactoferrin emulsions increase the stability during storage, and emulsions containing protein only show change in size during the storage of 2 weeks (from 209 nm to around 1000 nm) [38]. This result may be due to the electrostatic interactions of polysaccharides on the protein surface that inhibit protein aggregation.

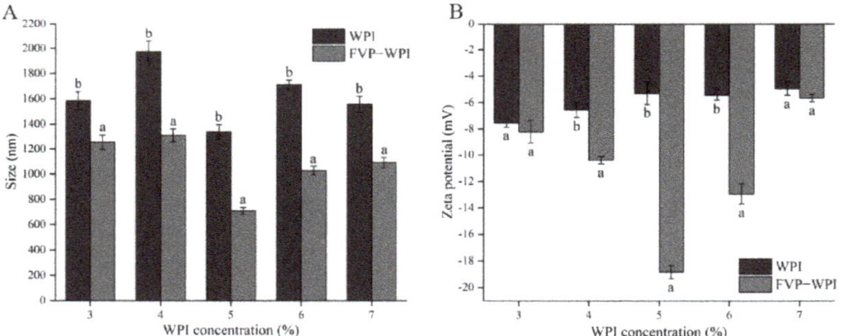

Figure 1. Characterization of whey protein isolate (WPI) and *Flammulina velutipes* polysaccharide (FVP)–WPI particles at pH 4.5. (**A**) Particle size; (**B**) zeta potential. Different letters in the figure represent significant differences in different samples of the same concentration ($p < 0.05$).

Zeta potential measurements are used to evaluate the stability of a dispersion system; the high absolute value of zeta potential indicates that there is a greater electrostatic repulsion between two molecules and the system is more stable. The stability of the system is affected by particle size and particle surface charge in the dispersed system [39]. The zeta potential of the investigated polysaccharide was −12.53 ± 0.51 mV. The zeta potential values of WPI and FVP–WPI complexes are shown in Figure 1B. These results showed that the noncovalent interactions between FVPs and WPI led to a decrease in the zeta potential of the solutions of the FVP–WPI complex. The zeta potentials of the WPI solutions ranged from -7.54 ± 0.31 mV at 3% WPI to -4.95 ± 0.49 mV at 7% WPI. In contrast, the zeta potential values of FVP–WPI complexes showed a low valley, decreasing from -8.22 ± 0.86 to -18.83 ± 0.50 mV as the WPI concentration increased from 3 to 5% since the charged area of the WPI surface was masked by FVPs as the complex formed. Interestingly, it has been previously reported that the zeta potential of protein decreases when interacting with polysaccharides [40,41]. Therefore, when the concentration of WPI increased, the surface potential of FVP–WPI decreased, and the number of FVP–WPI composite particles increased. However, the potential increased when the WPI concentration was 6 or 7%—this may be attributed to the surface of WPI that could not be covered by FVPs.

It could be observed in Figure 1 that when the concentration of WPI was 5%, the particle size and potential of FVP–WPI were the lowest. With the increase of WPI concentration, the particle size and potential of FVP–WPI increased. We confirmed that the best ratio of noncovalent interactions between FVPs and WPI was 1:12.5.

3.2. Turbiscan Measurements of Stability

Images of the WPI and FVP–WPI solutions stored from 0 to 15 d at 4 °C are presented in Figure 2. When the pH was adjusted to 4.5, precipitates were observed at the bottom of the cells for the WPI solutions at all assayed concentrations. The supernatant became more and more turbid with the increase of WPI concentrations. However, the solutions of FVP–WPI complex displayed better dispersal than the WPI solutions. No phase separation occurred in the FVP–WPI solutions after being stored for 15 d.

Figure 2. Images of WPI and FVP–WPI solutions over 15 d of storage.

Turbiscan is commonly used to predict and monitor the stability of systems [42]. Figures 3 and 4 show the BS of WPI and FVP–WPI solutions obtained at different times for the samples formulated from 3% WPI to 7% WPI via Turbiscan. All samples were measured on days 0, 5, 10, and 15. During this period, the FVP–WPI solutions showed better stabilities than the WPI solutions, and the destabilization of WPI solutions were caused by sedimentation. Sedimentation was evidenced by the increasing BS at the bottom zone (0–10 mm)—the range of length increased with the increasing of WPI concentrations. For all the solutions of FVP–WPI complex, the BS remained with no change over 15 d of storage, showing that no sedimentation occurred. In order to further verify the stability of FVP–WPI solutions, we carried out an accelerated stability test for 60 days according to a previously described method [43]. Images and BS of the FVP–WPI solutions stored from 0 to 60 d at 40 °C are presented in Figure S1. As a result, there was no phase separation in the heated FVP–WPI solutions during storage. The study of Yin et al. shows that soy polysaccharide and soy protein could form dispersible complexes at pH 3.25 in aqueous solution via electrostatic interactions [44]. These observations were consistent with the particle size and zeta potential analyses in Section 3.1, which indicated that the addition of FVPs can essentially maintain the stability of WPI at pH 4.5. The FVPs and WPI had opposite charges at pH 4.5, which caused the electrostatic attractions and drove the formation of complex through noncovalent interactions. We discovered that solution of FVP–WPI complex had higher storage stability than the WPI solutions.

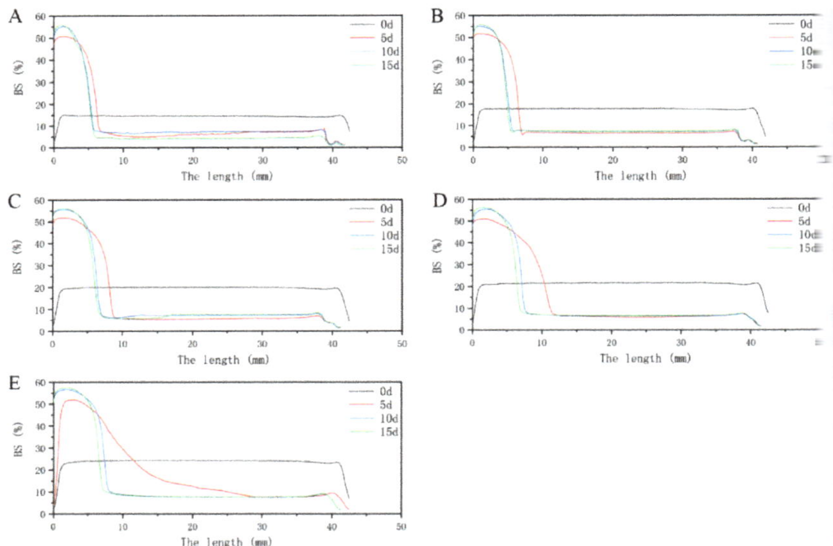

Figure 3. Back scattering (BS) of WPI solutions over storage time under static conditions. (**A**) 3% WPI solution; (**B**) 4% WPI solution; (**C**) 5% WPI solution; (**D**) 6% WPI solution; (**E**) 7% WPI solution.

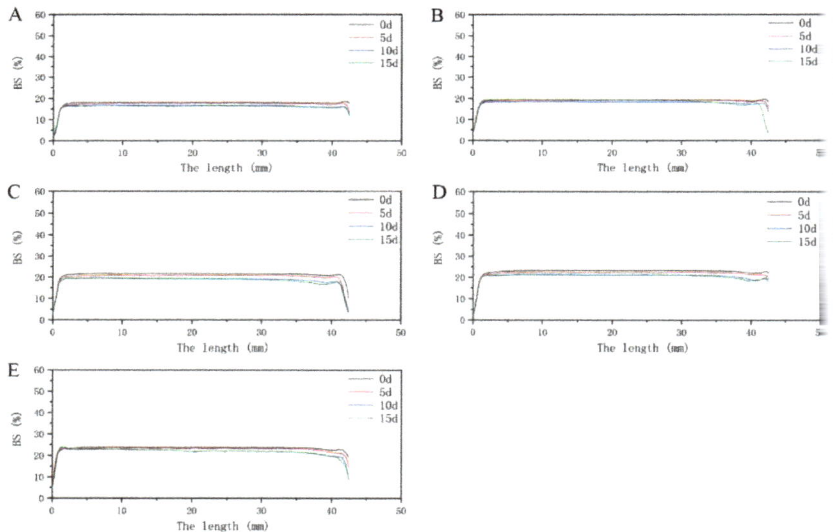

Figure 4. Back scattering (BS) of FVP–WPI solutions over storage time under static conditions. (**A**) 0.4% FVP–3% WPI solution; (**B**) 0.4% FVP–4% WPI solution; (**C**) 0.4% FVP–5% WPI solution; (**D**) 0.4% FVP–6% WPI solution; (**E**) 0.4% FVP–7% WPI solution.

3.3. Antioxidant Activities

The antioxidant activities of 0.4% FVPs and 0.4% FVP–5% WPI were assessed using three radicals, including ABTS, hydroxyl, and superoxide anion radicals (Figure 5A). FVP–WPI complexes displayed better ABTS radical scavenging activity than the FVPs, where the FVPs exhibited better antioxidant activity than the FVP–WPI solution with respect to its hydroxyl and superoxide anion radical scavenging activities. The difference of hydroxyl radical scavenging activities between FVP and FVP–WPI may result from the uronic acid in FVP, which was masked by WPI during interaction with WPI. Some studies reported that the content of uronic acid in polysaccharides is related to their radical scavenging activity [45,46]. The result of superoxide anion radical scavenging activities may result from the change in conformation; the result in Section 3.5 showed that the conformation of FVPs shifted when they interacted with WPI. The study of Zhang et al. shows that the biological activities of lentinan are correlated with the conformation of the molecule, and the biological activity decreases when the conformation is destroyed [46,47]. This will be of importance in designing a product that is stabilized by polysaccharide–protein complexes with objectives of delivering nutrients with regard to health benefits simultaneously.

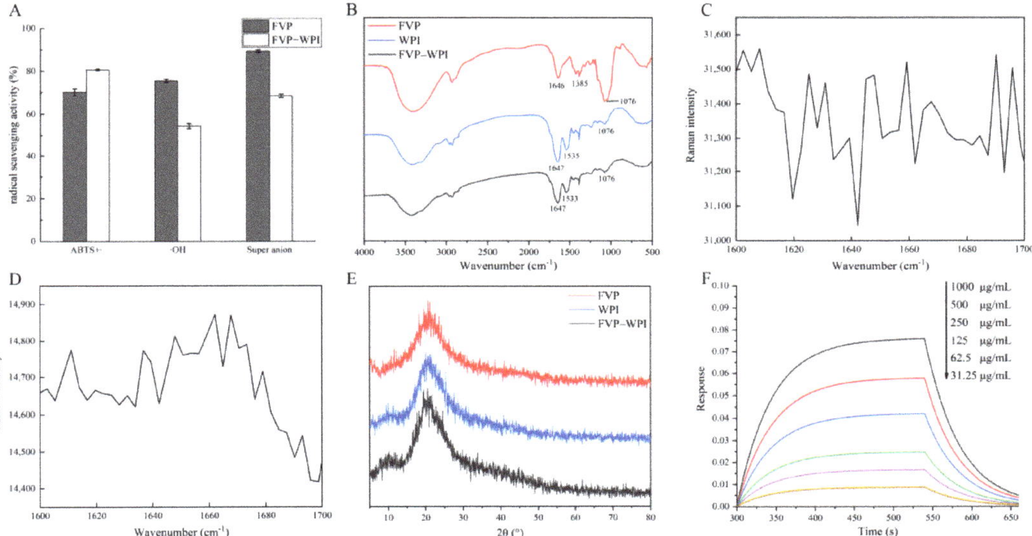

Figure 5. Antioxidant activities and structures of FVPs, WPI, and FVP–WPI. (**A**) The antioxidant activities of FVPs and FVP–WPI, (**B**) the FTIR spectra of the FVPs, WPI, and FVP–WPI complex, (**C,D**) the amide I bands of WPI and FVP–WPI in Raman spectra, (**E**) the XRD spectra of the WPI, FVPs, and FVP–WPI, (**F**) the fitting curves for association and dissociation of FVPs and WPI.

3.4. FTIR and Raman Spectroscopy

The FTIR spectra of the freeze-dried FVPs, WPI, and 0.4% FVP–5% WPI complex powders are presented in Figure 5B. In previous research, the strong band in the FVP spectrum at approximately 1076 cm^{-1} was ascribed to pyran-type structure, the absorption peak at approximately 1385 cm^{-1} was due to C–H deformation vibration, and the strong peak at approximately 1647 cm^{-1} was associated with C=O asymmetrical stretching vibrations [48,49]. In a previous study, the amide I and II peaks were related to the secondary structure, for instance, α-helix and β-sheet [50]. The amide I peak represented the C=O stretching vibrations, and the amide II peak was due to the N=H bending and C=N stretching vibrations. Our results showed that the amide I peak of WPI was observed at 1647 cm^{-1}, and the amide II peak was observed at 1535 cm^{-1}, with one band detected at 1076 cm^{-1}. In addition, the amide I peak of the FVP–WPI spectrum was at 1647 cm^{-1}, the amide II peak shifted to 1533 cm^{-1} compared with that observed in WPI, and a band at 1076 cm^{-1} was also observed. These results revealed that addition of FVPs had no clear effects on the positions of amide I and II peaks of WPI. Furthermore, we estimated the changes of protein secondary structure quantitatively through Raman spectra.

Changes of the position and intensity of the spectral peaks in Raman spectra reflect changes in protein structures [30]. Conformations of the primary protein chain were determined by the characteristic peak of amide I bands (1600–1700 cm^{-1}), and the Raman spectra of WPI and 0.4% FVP–5% WPI are shown in Figure 5C,D. Fitting of the amide I band was performed typically to estimate protein secondary structure quantitatively, and the percentages of α-helix, β-sheet, β-turn, random coil, and amino acid side chains were calculated by the relative contributions of respective components. The frequences and corresponding percentages of each secondary structure for the two samples obtained by performing fitting calculations are shown in Table 2. After noncovalent interactions, the α-helix, β-turn, and random coil contents were found to be higher in the FVP–WPI complex than in WPI ($p < 0.05$). Therefore, formation of the FVP–WPI complexes appeared to change the spatial secondary structure of WPI and promote the formation of α-helices, β-turns, and random coils in the FVP–WPI complexes. In our work, the variation trends of percentages of each secondary structure were different from those reported by Zhang et al., which may have been caused by diverse interactions or diverse polysaccharides [51]. We found the secondary structures of WPI were changed significantly after interacting with FVPs, including α-helix, β-turn, and random coil.

Table 2. Content of secondary structural components of WPI and FVP–WPI.

		Structural Contribution (%)	
Structure	**Frequency (cm^{-1})**	**WPI**	**FVP–WPI**
α-Helix	1654	24.85 ± 0.22 [b]	32.59 ± 0.38 [a]
β-Sheet	1667 and 1676	46.49 ± 0.45 [a]	31.70 ± 0.36 [b]
β-Turn	1632 and 1684	11.29 ± 0.67 [b]	16.08 ± 0.76 [a]
Random coil	1643	10.87 ± 0.63 [b]	14.85 ± 0.4 [a]
Amino acid side	1614	4.53 ± 0.67 [a]	4.54 ± 0.27 [a]

Error bars indicate mean values ± standard deviations ($n = 3$), Different superscripts within a line indicate significant differences ($p < 0.05$).

3.5. XRD Analysis

XRD analyses were performed to obtain crystal structures. The XRD spectra of the freeze-dried FVPs, WPI, and 0.4% FVP–5% WPI complex powders are presented in Figure 5E. Two peaks were observed at the reflections of approximately 10° and 21° in the WPI and FVP–WPI samples, while only one peak appeared in the spectrum of FVPs at a reflection of approximately 21°. The crystallinity of FVP–WPI was lower than FVPs or WPI, as there was noncovalent interactions with FVPs, indicating that the crystal became more amorphous, which was similar to the work of Xu et al. [52]. This result was due to the FVPs entering the crystalline structure of WPI, which resulted in the amorphous structure of the FVP–WPI complex.

3.6. Assessment of the FVP–WPI Interaction by BLI

The kinetics assay was performed using five sections, including two baselines (PBS and PBST), a loading of ligand (WPI), an association of the analyte (FVPs), and a dissociation step in PBST. We analyzed association and dissociation mainly as a result of the three sections (two baselines and a loading of ligand) in different concentrations assay, which were found to be consistent. The fitting curves for association and dissociation are shown in Figure 5F, and the R^2 value of the fitting curves was 0.99. In the gradient test, the FVPs exhibited a gradient interaction with the receptor protein WPI, and the equilibrium dissociation constant (K_D) between WPI and FVPs was calculated as 1.736×10^{-4} M, suggesting that there was an interaction that occurred between FVPs and WPI. Similarly, Wallner recently showed in protein/liposome binding interactions based on the BLI that liposome formed stable complexes with the protein, wherein the K_D between protein and liposome was calculated as 5.166×10^{-5} M–1.845×10^{-4} M, which was different from our study [53]. This difference may be due to bond strength of materials. We believe that BLI can be used to measure the kinetics in interactions between polysaccharides and proteins.

3.7. In Vitro Digestibility

A double digestion process in the proteomics analysis was used to produce peptides, after which the protein undergoing transformations at the peptide level were determined. The volcano plots were used to show the different peptides between FVP–WPI and WPI (the essence of a volcano plot is a scatter plot). The y-axis represented $-\log_{10}$ (p-value). The higher the y-coordinate value, the smaller the p-value, that is, the more significant. The x-axis represented difference (FVP–WPI vs. WPI). The positive x-coordinate value represented the fact that the peptide abundance of FVP–WPI was higher than the peptide abundance of WPI. The negative x-coordinate value represented the fact that the peptide abundance of FVP–WPI was smaller than the peptide abundance of WPI. Volcano plots of the 0.4% FVP–5% WPI complex compared with WPI are shown in Figure 6, where limited digestion by pepsin (non-specific protease) was performed (10 or 120 min), followed by complete proteolysis using trypsin (specific protease). After decompositions via pepsin and trypsin digestion, three types of peptides were generated, including peptides both having lysine (Lys, K) and arginine (Arg, R) terminals, peptides that have K or R terminals, and other peptides (trypsin used in our study is a serine protease that specifically cleaves at the carboxyl side of lysine and arginine residues). After 10 or 120 min of simulated gastric fluid digestion, more peptides that had K or R terminals were observed in the FVP–WPI group compared with the WPI group. This finding was consistent with those of previous studies [54]. This was attributed to the protection offered by FVPs, which could make the WPI keep away from gastric fluid digestion. Meanwhile, the trypsin digested samples drastically, and thus there were more peptides that had K or R terminals in the system. Abundance of peptides in the FVP–WPI group increased significantly after 120 min of digestion compared with that observed on minute 10 (Figure 6B). This was due to long-time digestion by gastric fluid. These results indicated that the formation of the FVP–WPI complexes protected WPI from digestion via pepsin, which also can increase satiety of beverages that contain FVP–WPI. On the basis of the nutrition and function of the complex, the highly stable aqueous solution produced using the complex will have broad application prospects in the future.

3.8. 3D Structures of Potential Binding Regions

The potential binding regions were inferred by peptide abundance changes calculated via proteomics data and were mapped onto WPI sequences. Software tools such as Pymol can be used to map MS data to the protein structure to achieve data visualization [55]. The potential binding regions are shown as 3D structures in Figure 7. As shown in Figure 7, the peptide fragment "VGINYWLAHK" was probably a binding region between FVPs and alpha-lactalbumin; the peptide fragment "TKIPAVFKIDALNENK" or "IDAL-NENKVLVLDTDYKK" was likely docked with β-lactoglobulin. The peptides TPVSEKVTK and DAFLGSFLYEYSR were the binding region of serum albumin to FVPs. FVPs interacted with lactoperoxidase through binding with peptide LFQPTHK and DGGIDPLVR, and the peptide fragments "KANEGLTWNSLK" and "APVDAFK" were the binding region of lactotransferrin to FVPs. The previous study showed that the peptide fragment "VGINYWLAHK" was one of the binding regions of alpha-lactalbumin to *Tremella fuciformis* polysaccharide [41]. Accurate binding region of WPI and polysaccharide would be inferred through a mass of dates in further research. In this study, results about the predictions of binding regions implied that the proteomics is a useful technique that will be further developed for studying the interaction properties between polysaccharide and protein in the future.

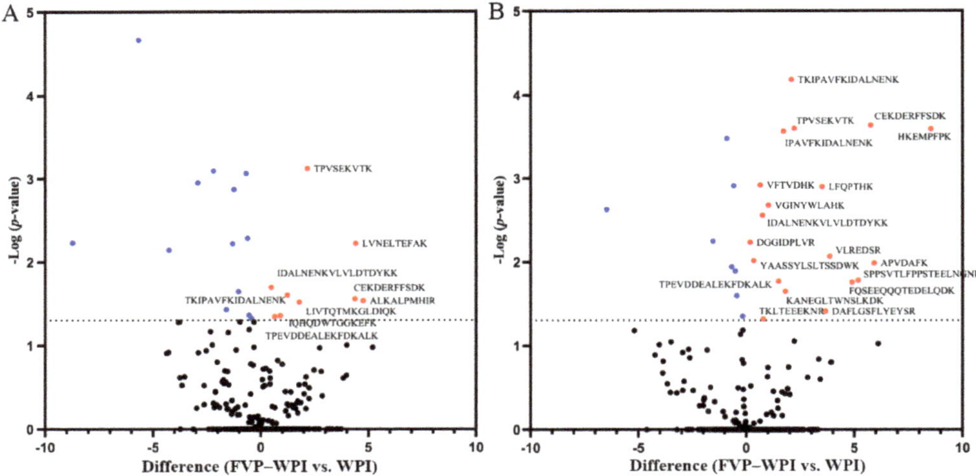

Figure 6. The volcano plots of the change in peptide abundance between FVP–WPI and WPI (control). (**A**) Differences of peptides digested for 10 min. (**B**) Differences of peptides digested for 120 min. Significant changes (*p*-value < 0.05) appear in the upper left (significant decrease) and upper right quadrants (significant increase) of the plot. The color is used to distinguish whether the sequences are different. Red circles represent significant increase and blue circles represent significant decrease.

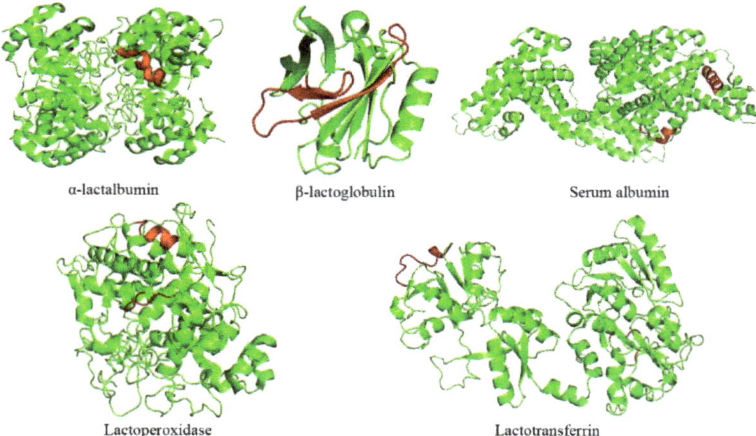

Figure 7. 3D structure of potential binding regions in WPI (red regions indicate the potential binding regions with FVPs).

4. Conclusions

The FVP–WPI complexes were prepared via noncovalent interactions of FVP and WPI. The results of our work showed that the FVP–WPI complexes dispersed well in an aqueous solution at pH 4.5, had high stability, and maintained antioxidant activities. The results of FTIR, Raman, and XRD primarily showed the differences in secondary structures and conformation between FVP–WPI and WPI. In addition, we confirmed that FVPs did form stable complexes with WPI through noncovalent bonding according to kinetics via BLI. Proteomics analysis proved that FVPs can protect WPI from gastric digestion, and inferred the binding site between FVPs and WPI, which further proved that FVPs can form stable complexes with WPI through noncovalent interactions. These results all proved that FVPs can form stable complexes with WPI at pH 4.5 and showed that FVPs are a great material to change the stability of WPI. Our results provide a foundation for the application of FVPs and WPI to develop the stability and functional properties of WPI beverages in the food industry.

Supplementary Materials: The following are available online at https://www.mdpi.com/2304-815 8/10/1/1/s1, Figure S1: Images and back scattering (BS) of FVP–WPI solutions over 60 d of storage.

Author Contributions: J.S., L.Z., and N.L. conceived and designed the experiments; R.J., H.L., and Y.X. contributed reagents, materials, and analytical tools and participated in the experiments; J.S., M.L., and X.T. analyzed the data; and J.S., M.L., L.Z., and N.L. reviewed and edited the manuscript. All authors have read and agreed to the published version of the manuscript.

Funding: This research was funded by the National Key Research and Development Program of China (no. 2018YFC1604304).

Acknowledgments: We thank Key Lab of Meat Processing and Quality Control for the stability analysis.

Conflicts of Interest: There are no conflict of interest to declare.

References

1. Chanet, A.; Verlaan, S.; Salles, J.; Giraudet, C.; Patrac, V.; Pidou, V.; Pouyet, C.; Hafnaoui, N.; Blot, A.; Cano, N.; et al. Supplementing breakfast with a vitamin d and leucine-enriched whey protein medical nutrition drink enhances postprandial muscle protein synthesis and muscle mass in healthy older men. *J. Nutr.* **2017**, *147*, 2262–2271. [CrossRef] [PubMed]

2. Martin-Rincon, M.; Perez-Suarez, I.; Pérez-López, A.; Ponce-González, J.G.; Morales-Alamo, D.; de Pablos-Velasco, P.; Holmberg, H.C.; Calbet, J.A.L. Protein synthesis signaling in skeletal muscle is refractory to whey protein ingestion during a severe energy deficit evoked by prolonged exercise and caloric restriction. *Int. J. Obes.* **2019**, *43*, 872–882. [CrossRef] [PubMed]

3. Wang, W.-Q.; Zhang, L.-W.; Han, X.; Lu, Y. Cheese whey protein recovery by ultrafiltration through transglutaminase (TG) catalysis whey protein cross-linking. *Food Chem.* **2017**, *215*, 31–40. [CrossRef]

4. Liu, J.; Wang, X.; Zhao, Z. Effect of whey protein hydrolysates with different molecular weight on fatigue induced by swimming exercise in mice. *J. Sci. Food Agric.* **2014**, *94*, 126–130. [CrossRef] [PubMed]

5. Dale, M.J.; Thomson, R.L.; Coates, A.M.; Howe, P.R.C.; Brown, A.; Buckley, J.D. Protein hydrolysates and recovery of muscle damage following eccentric exercise. *Funct. Foods Health Dis.* **2015**, *5*. [CrossRef]

6. Beecher, J.W.; Drake, M.A.; Luck, P.J.; Foegeding, E.A. Factors regulating astringency of whey protein beverages. *J. Dairy Sci.* **2008**, *91*, 2553–2560. [CrossRef]

7. Dai, Q.; Zhu, X.; Abbas, S.; Karangwa, E.; Zhang, X.; Xia, S.; Feng, B.; Jia, C. Stable nanoparticles prepared by heating electrostatic complexes of whey protein isolate-dextran conjugate and chondroitin sulfate. *J. Agric. Food Chem.* **2015**, *63*, 4179–4189. [CrossRef]

8. Vardhanabhuti, B.; Yucel, U.; Coupland, J.N.; Foegeding, E.A. Interactions between β-lactoglobulin and dextran sulfate at near neutral pH and their effect on thermal stability. *Food Hydrocoll.* **2009**, *23*, 1511–1520. [CrossRef]

9. Taherian, A.R.; Britten, M.; Sabik, H.; Fustier, P. Ability of whey protein isolate and/or fish gelatin to inhibit physical separation and lipid oxidation in fish oil-in-water beverage emulsion. *Food Hydrocoll.* **2011**, *25*, 868–878. [CrossRef]

10. Klein, M.; Aserin, A.; Svitov, I.; Garti, N. Enhanced stabilization of cloudy emulsions with gum Arabic and whey protein isolate. *Colloids Surf. Biointerfaces* **2010**, *77*, 75–81. [CrossRef]

11. Mao, L.; Boiteux, L.; Roos, Y.H.; Miao, S. Evaluation of volatile characteristics in whey protein isolate–pectin mixed layer emulsions under different environmental conditions. *Food Hydrocoll.* **2014**, *41*, 79–85. [CrossRef]

12. Qi, P.X.; Xiao, Y.; Wickham, E.D. Stabilization of whey protein isolate (WPI) through interactions with sugar beet pectin (SBP) induced by controlled dry-heating. *Food Hydrocoll.* **2017**, *67*, 1–13. [CrossRef]

13. Tuohy, K.M.; Hinton, D.J.; Davies, S.J.; Crabbe, M.J.; Gibson, G.R.; Ames, J.M. Metabolism of Maillard reaction products by the human gut microbiota—implications for health. *Mol. Nutr. Food Res.* **2006**, *50*, 847–857. [CrossRef] [PubMed]

14. ALjahdali, N.; Carbonero, F. Impact of Maillard reaction products on nutrition and health: Current knowledge and need to understand their fate in the human digestive system. *Crit. Rev. Food Sci. Nutr.* **2019**, *59*, 474–487. [CrossRef]

15. Rodríguez Patino, J.M.; Pilosof, A.M.R. Protein–Polysaccharide interactions at fluid interfaces. *Food Hydrocoll.* **2011**, *25*, 1925–1937. [CrossRef]

16. Xin, X.; Zheng, K.; Niu, Y.; Song, M.; Kang, W. Effect of *Flammulina velutipes* (golden needle mushroom, eno-kitake) polysaccharides on constipation. *Open Chem.* **2018**, *16*, 155–162. [CrossRef]

17. Liu, Y.; Zhang, B.; Ibrahim, S.A.; Gao, S.S.; Yang, H.; Huang, W. Purification, characterization and antioxidant activity of polysaccharides from *Flammulina velutipes* residue. *Carbohydr. Polym.* **2016**, *145*, 71–77. [CrossRef]

18. Meng, Y.; Yan, J.; Yang, G.; Han, Z.; Tai, G.; Cheng, H.; Zhou, Y. Structural characterization and macrophage activation of a hetero-galactan isolated from *Flammulina velutipes*. *Carbohydr. Polym.* **2018**, *183*, 207–218. [CrossRef]
19. Zhang, T.; Ye, J.; Xue, C.; Wang, Y.; Liao, W.; Mao, L.; Yuan, M.; Lian, S. Structural characteristics and bioactive properties of a novel polysaccharide from *Flammulina velutipes*. *Carbohydr. Polym.* **2018**, *197*, 147–156. [CrossRef]
20. Zhang, Y.; Li, H.; Hu, T.; Li, H.; Jin, G.; Zhang, Y. Metabonomic profiling in study hepatoprotective effect of polysaccharides from *Flammulina velutipes* on carbon tetrachloride-induced acute liver injury rats using GC-MS. *Int. J. Biol. Macromol.* **2018**, *113*, 285–293. [CrossRef]
21. Zhuang, X.; Wang, L.; Jiang, X.; Chen, Y.; Zhou, G. The effects of three polysaccharides on the gelation properties of myofibrillar protein: Phase behaviour and moisture stability. *Meat Sci.* **2020**, *170*, 108228. [CrossRef] [PubMed]
22. Wu, F.; Zhou, C.; Zhou, D.; Ou, S.; Huang, H. Structural characterization of a novel polysaccharide fraction from *Hericium erinaceus* and its signaling pathways involved in macrophage immunomodulatory activity. *J. Funct. Foods* **2017**, *37*, 574–585. [CrossRef]
23. Zhang, S.J.; Hu, T.T.; Chen, Y.Y.; Wang, S.; Kang, Y.F. Analysis of the polysaccharide fractions isolated from pea (*Pisum sativum* L.) at different levels of purification. *J. Food Biochem.* **2020**, *44*, e13248. [CrossRef] [PubMed]
24. Guzey, D.; McClements, D.J. Characterization of β-lactoglobulin–chitosan interactions in aqueous solutions: A calorimetry, light scattering, electrophoretic mobility and solubility study. *Food Hydrocoll.* **2006**, *20*, 124–131. [CrossRef]
25. Liu, Q.; Jing, Y.; Han, C.; Zhang, H.; Tian, Y. Encapsulation of curcumin in zein/caseinate/sodium alginate nanoparticles with improved physicochemical and controlled release properties. *Food Hydrocoll.* **2019**, *93*, 432–442. [CrossRef]
26. Montes de Oca-Avalos, J.M.; Candal, R.J.; Herrera, M.L. Colloidal properties of sodium caseinate-stabilized nanoemulsions prepared by a combination of a high-energy homogenization and evaporative ripening methods. *Food Res. Int.* **2017**, *100*, 143–150. [CrossRef] [PubMed]
27. Chen, G.; Fang, C.; Ran, C.; Tan, Y.; Yu, Q.; Kan, J. Comparison of different extraction methods for polysaccharides from bamboo shoots (*Chimonobambusa quadrangularis*) processing by-products. *Int. J. Biol. Macromol.* **2019**, *130*, 903–914. [CrossRef]
28. Koh, H.S.A.; Lu, J.; Zhou, W. Structure characterization and antioxidant activity of fucoidan isolated from *Undaria pinnatifida* grown in New Zealand. *Carbohydr. Polym.* **2019**, *212*, 178–185. [CrossRef]
29. Yan, J.; Zhu, L.; Qu, Y.; Qu, X.; Mu, M.; Zhang, M.; Muneer, G.; Zhou, Y.; Sun, L. Analyses of active antioxidant polysaccharides from four edible mushrooms. *Int. J. Biol. Macromol.* **2019**, *123*, 945–956. [CrossRef]
30. Zhuang, X.; Han, M.; Bai, Y.; Liu, Y.; Xing, L.; Xu, X.-L.; Zhou, G.-H. Insight into the mechanism of myofibrillar protein gel improved by insoluble dietary fiber. *Food Hydrocoll.* **2018**, *74*, 219–226. [CrossRef]
31. Lu, X.; Shi, C.; Zhu, J.; Li, Y.; Huang, Q. Structure of starch-fatty acid complexes produced via hydrothermal treatment. *Food Hydrocoll.* **2019**, *88*, 58–67. [CrossRef]
32. Stengel, K.F.; Harden-Bowles, K.; Yu, X.; Rouge, L.; Yin, J.; Comps-Agrar, L.; Wiesmann, C.; Bazan, J.F.; Eaton, D.L.; Grogan, J.L. Structure of TIGIT immunoreceptor bound to poliovirus receptor reveals a cell-cell adhesion and signaling mechanism that requires *cis-trans* receptor clustering. *Proc. Natl. Acad. Sci. USA* **2012**, *109*, 5399–5404. [CrossRef] [PubMed]
33. Dai, J.; Huang, Y.J.; He, X.; Zhao, M.; Wang, X.; Liu, Z.S.; Xue, W.; Cai, H.; Zhan, X.Y.; Huang, S.Y.; et al. Acetylation blocks cGAS activity and inhibits self-DNA-induced autoimmunity. *Cell* **2019**, *176*, 1447–1460.e14. [CrossRef] [PubMed]
34. Zhang, X.R.; Qi, C.H.; Cheng, J.P.; Liu, G.; Huang, L.J.; Wang, Z.F.; Zhou, W.X.; Zhang, Y.X. *Lycium barbarum* polysaccharide LBPF4-OL may be a new Toll-like receptor 4/MD2-MAPK signaling pathway activator and inducer. *Int. Immunopharmacol.* **2014**, *19*, 132–141. [CrossRef] [PubMed]
35. Zhang, Y.H.; Shetty, K.; Surleac, M.D.; Petrescu, A.J.; Schatz, D.G. Mapping and quantitation of the interaction between the recombination activating gene proteins RAG1 and RAG2. *J. Biol. Chem.* **2015**, *290*, 11802–11817. [CrossRef]
36. Akl, M.A.; Kartal-Hodzic, A.; Oksanen, T.; Ismael, H.R.; Afouna, M.M.; Yliperttula, M.; Samy, A.M.; Viitala, T. Factorial design formulation optimization and in vitro characterization of curcumin-loaded PLGA nanoparticles for colon delivery. *J. Drug Deliv. Sci. Technol.* **2016**, *32*, 10–20. [CrossRef]
37. Zhang, L.; Boeren, S.; Smits, M.; van Hooijdonk, T.; Vervoort, J.; Hettinga, K. Proteomic study on the stability of proteins in bovine, camel, and caprine milk sera after processing. *Food Res. Int.* **2016**, *82*, 104–111. [CrossRef]
38. Zhao, J.; Wei, T.; Wei, Z.; Yuan, F.; Gao, Y. Influence of soybean soluble polysaccharides and beet pectin on the physicochemical properties of lactoferrin-coated orange oil emulsion. *Food Hydrocoll.* **2015**, *44*, 443–452. [CrossRef]
39. Kim, E.-A.; Kim, J.-Y.; Chung, H.-J.; Lim, S.-T. Preparation of aqueous dispersions of coenzyme Q10 nanoparticles with amylomaize starch and its dextrin. *LWT* **2012**, *47*, 493–499. [CrossRef]
40. Liu, J.; Shim, Y.Y.; Shen, J.; Wang, Y.; Reaney, M.J.T. Whey protein isolate and flaxseed (*Linum usitatissimum* L.) gum electrostatic coacervates: Turbidity and rheology. *Food Hydrocoll.* **2017**, *64*, 18–27. [CrossRef]
41. Hu, J.; Zhao, T.; Li, S.; Wang, Z.; Wen, C.; Wang, H.; Yu, C.; Ji, C. Stability, microstructure, and digestibility of whey protein isolate—*Tremella fuciformis* polysaccharide complexes. *Food Hydrocoll.* **2019**, *89*, 379–385. [CrossRef]
42. Adeyi, O.; Ikhu-Omoregbe, D.; Jideani, V. Emulsion stability and steady shear characteristics of concentrated oil-in-water emulsion stabilized by gelatinized bambara groundnut flour. *Asian J. Chem.* **2014**, *26*, 4995–5002. [CrossRef]
43. Niu, B.; Shao, P.; Feng, S.; Qiu, D.; Sun, P. Rheological aspects in fabricating pullulan-whey protein isolate emulsion suitable for electrospraying: Application in improving β-carotene stability. *LWT* **2020**, *129*. [CrossRef]
44. Yin, B.; Deng, W.; Xu, K.; Huang, L.; Yao, P. Stable nano-sized emulsions produced from soy protein and soy polysaccharide complexes. *J. Colloid Interface Sci.* **2012**, *380*, 51–59. [CrossRef]

45. Chen, H.X.; Zhang, M.; Xie, B.J. Quantification of uronic acids in tea polysaccharide conjugates and their antioxidant properties. *J. Agric. Food Chem.* **2004**, *52*, 3333–3336. [CrossRef]
46. Sun, Y.-X.; Liu, J.-C.; Kennedy, J.F. Purification, composition analysis and antioxidant activity of different polysaccharide conjugates (APPs) from the fruiting bodies of *Auricularia polytricha*. *Carbohydr. Polym.* **2010**, *82*, 299–304. [CrossRef]
47. Zhang, L.; Li, X.; Xu, X.; Zeng, F. Correlation between antitumor activity, molecular weight, and conformation of lentinan. *Carbohydr. Res.* **2005**, *340*, 1515–1521. [CrossRef]
48. Chen, G.; Li, C.; Wang, S.; Mei, X.; Zhang, H.; Kan, J. Characterization of physicochemical properties and antioxidant activity of polysaccharides from shoot residues of bamboo (*Chimonobambusa quadrangularis*): Effect of drying procedures. *Food Chem.* **2019**, *292*, 281–293. [CrossRef]
49. Rong, Y.; Yang, R.; Yang, Y.; Wen, Y.; Liu, S.; Li, C.; Hu, Z.; Cheng, X.; Li, W. Structural characterization of an active polysaccharide of longan and evaluation of immunological activity. *Carbohydr. Polym.* **2019**, *213*, 247–256. [CrossRef]
50. Jia, Z.; Zheng, M.; Tao, F.; Chen, W.; Huang, G.; Jiang, J. Effect of covalent modification by (−)-epigallocatechin-3-gallate on physicochemical and functional properties of whey protein isolate. *LWT* **2016**, *66*, 305–310. [CrossRef]
51. Zhang, S.; Zhang, Z.; Lin, M.; Vardhanabhuti, B. Raman spectroscopic characterization of structural changes in heated whey protein isolate upon soluble complex formation with pectin at near neutral pH. *J. Agric. Food Chem.* **2012**, *60*, 12029–12035. [CrossRef] [PubMed]
52. Xu, Z.; Hao, N.; Li, L.; Zhang, Y.; Yu, L.; Jiang, L.; Sui, X. Valorization of soy whey wastewater: How epigallocatechin-3-gallate regulates protein precipitation. *ACS Sustain. Chem. Eng.* **2019**, *7*, 15504–15513. [CrossRef]
53. Wallner, J.; Lhota, G.; Jeschek, D.; Mader, A.; Vorauer-Uhl, K. Application of Bio-Layer Interferometry for the analysis of protein/liposome interactions. *J. Pharm. Biomed. Anal.* **2013**, *72*, 150–154. [CrossRef] [PubMed]
54. Zhang, S.; Vardhanabhuti, B. Intragastric gelation of whey protein-pectin alters the digestibility of whey protein during in vitro pepsin digestion. *Food Funct.* **2014**, *5*, 102–110. [CrossRef]
55. Schopper, S.; Kahraman, A.; Leuenberger, P.; Feng, Y.; Piazza, I.; Muller, O.; Boersema, P.J.; Picotti, P. Measuring protein structural changes on a proteome-wide scale using limited proteolysis-coupled mass spectrometry. *Nat. Protoc.* **2017**, *12*, 2391–2410. [CrossRef]

Article

The Impact of Different Factors on the Quality and Volatile Organic Compounds Profile in "Bryndza" Cheese

Jana Štefániková [1,*], **Viera Ducková** [2], **Michal Miškeje** [1], **Miroslava Kačániová** [3,4] and **Margita Čanigová** [2]

[1] AgroBioTech Research Centre, Slovak University of Agriculture in Nitra, Tr. A. Hlinku 2, 949 76 Nitra, Slovakia; michal.miskeje@uniag.sk

[2] Department of Technology and Quality of Animal Products, Faculty of Biotechnology and Food Sciences, Slovak University of Agriculture in Nitra, Tr. A. Hlinku 2, 949 76 Nitra, Slovakia; viera.duckova@uniag.sk (V.D.); margita.canigova@uniag.sk (M.Č.)

[3] Department of Fruit Science, Viticulture and Enology, Faculty of Horticulture and Landscape Engineering, Slovak University of Agriculture in Nitra, Tr. A. Hlinku 2, 949 76 Nitra, Slovakia; miroslava.kacaniova@gmail.com

[4] Department of Bioenergy, Food Technology and Microbiology, Institute of Food Technology and Nutrition, University of Rzeszow, Cwiklinkiej 1, 35601 Rzeszow, Poland

* Correspondence: jana.stefanikova@uniag.sk; Tel.: +421-376-414-911

Received: 31 July 2020; Accepted: 27 August 2020; Published: 29 August 2020

Abstract: The aim of this study was to evaluate the influence of different factors on the basic physicochemical and microbiological parameters, as well as volatile organic compounds of traditionally (farm) and industrially produced "bryndza" cheese. The samples were obtained from eight producers in different areas of Slovakia during the ewe's milk production season, from May to September. The physicochemical parameters set by the legislation were monitored by reference methods. The "bryndza" cheese microbiota was determined by using the plate cultivation method. There was analysis of volatile organic compounds carried out by electronic nose, as well as gas chromatography mass spectrometry. Seasonality and production technology (traditional and industrial ones) are the main factors that affect the standard quality of "bryndza" cheese. Lactic acid bacteria were dominated from bacterial microbiota, mostly presumptive lactococci, followed presumptive lactobacilli and enterococci. The numbers of coliform bacteria were higher in traditionally produced "bryndza" cheese than in industrially produced "bryndza" cheese. The presence of *Dipodascus geotrichum* was detected in all samples. There were key volatile organic compounds such as ethyl acetate, isoamyl acetate, 2-butanone, hexanoic acid, D-limonene, and 2,3-butanedione. The statistically significant differences were found among "bryndza" cheese samples and these differences were connected with the type of milk and dairies.

Keywords: "bryndza" cheese; electronic nose; gas chromatography; volatile organic compounds; microbiota

---

1. Introduction

The traditional Slovak ewe's milk product is "bryndza" cheese or local cheese "oštiepok" [1,2]. Slovak "bryndza" cheese is natural, white, spreadable cheese, and manufactured by the traditional method. It is recognized in the European Union by Protected Geographic Indication (PGI) status as cheese produced in specified mountainous regions of Slovakia [3], where unpasteurized ewe's milk is processed immediately after milking by renneting at 29–31 °C for 30 min, using chymosin or chymosin–identical rennet [4], and the cheese grain is formed into the lump cheese. The lump cheese is drained at 18–22 °C for 24 h and is left to ripen for 3 days at 18–22 °C. The production process continues by further ripening at 12–15 °C for 7–10 days. Ripened ewe's lump cheese is processed by removing the crust, pressed to remove whey and milled with salt solution (4–6% *w/w*), in order to obtain the specific creamy texture of "bryndza". The mountainous regions of Slovakia differ in altitude, climate, geological, and vegetation profile and there are some scientific evidences about variability of "bryndza" cheese. There are several studies related to the variability of "bryndza" that are lack of common characteristics of this Slovak cheese [5]. Ewe's cheese represents a matrix with a specific composition which reflected ewe's milk, as well as different autochthonous non-starter lactic acid bacteria (NSLAB) that produce typical aroma profile of ewe's lump cheese, barrelled ewe's cheese and "bryndza" cheese [4,5]. Kacaniova et al. [6] identified 870 isolates from coliforms, enterococci, lactic acid bacteria (LAB) and yeasts in Slovak "bryndza" cheese by MALDI–TOF MS profiling. The most frequently identified species of gram-negative bacteria were *Hafnia alvei* and *Klebsiella oxytoca*. The most frequently identified species of gram-positive bacteria were *Lactococcus lactis* and *Lactobacillus paracasei*. LAB group was represented by *Lactobacillus*, *Lactococcus* and *Pediococcus*. Pangallo et al. [7] confirmed that microorganisms that belong to the species *Galactomyces candidus* and *Yarrowia lipolitica* as typical yeasts of "bryndza" cheese.

Numerous volatile organic compounds (VOCs) of cheese, including raw milk-based ewe's cheese are formed by proteolysis and by the subsequent transformation of amino acids [8] to α-keto acids [9]. There has been two different major pathways of amino acid degradation identified in *L. lactis* [10] so far. The first pathway is initiated by an elimination reaction of methionine catalyzed by amino acid lyases, and leading to major sulfur aroma compounds [11,12]. The second pathway is initiated by a transamination reaction catalyzed by aminotransferases and resulting especially in volatile amino acids, branched chain amino acids and methionine [13,14]. The resulting α-ketoacids are then degraded to aldehydes, alcohols, carboxylic acids, esters, methanethiol and other sulfur compounds. The most of these compounds are produced by enzymatic degradation, but some of them are the result of chemical degradation in oxidation [15,16]. The characteristics of cheese (flavor, texture, and color) are influenced by the free fatty acids (FFAs) of the milk, the main products of enzymatic hydrolysis of triacylglycerides by esterases and lipases [17]. The goat and ewe's milk contain high levels of short-and medium-chain fatty acids, in comparison with cow's milk. Specifically, short-chain fatty acids are very important because of their low perception thresholds. Some studies point out that milk fatty acid content could be a distinguishing clue between breeds [18]. Partial lipolysis occurs in "bryndza" cheese during its ripening. Additionally, FFAs may participate in catabolic reactions and cause an increase in the amount of aroma compounds such as methyl ketones, esters, alkanes, lactones, aldehydes, and secondary alcohols [19].

VOCs are usually analyzed by gas chromatography (GC) after the extraction or pre-concentration of the volatile fraction. The most exhaustive methods for this purpose are high vacuum distillation (HVT), solvent-assisted aroma evaporation (SAFE) or solid phase microextraction (SPME) [20,21] combined with headspace. Sádecká et al. [20] used SPME with gas chromatography-olfactometry (GC–O) for the determination of volatile odorants in May "bryndza" cheese. There were 25 olfactometric responses from the groups of alcohols, aldehydes, esters, ketones, fatty acids, and hydrocarbons recorded, depending on the degree of cheese maturation and from a GC-O point of view. There was an electronic nose (e-nose) based on a gas chromatography used for aroma profile determination of soft, steamed cheese called "parenica" from Slovakia in a previous study [22].

The aim of this study was to obtain microbiological parameters and parallel information of principal VOCs in "bryndza" cheese produced by 8 Slovak producers with the use of e-nose and a GC-MS. The other goal was to prove a significant impact of the dairies and the type of milk on the content of VOCs in "bryndza" cheese.

2. Materials and Methods

2.1. Bryndza Cheese Samples

The samples of "bryndza" cheese were obtained from 8 different producers in Slovakia (B1–8) and detailed characteristics were got from the packages (Table 1). Bryndza cheese samples B1, B2, B4, B6, and B7 were produced in the farm dairies and the samples B3, B5, and B8 were produced in the industrial dairies. All samples were collected monthly from May 2019 to September 2019. Each sample (500 g) was transported to the laboratory at a temperature of 6 °C. There were physicochemical parameters, microbiological quality and volatile organic compounds within one day after the delivery determined in the representative samples. A total of 40 samples were analyzed.

Table 1. Characterization of analyzed bryndza cheese samples.

	B1	B2	B3	B4	B5	B6	B7	B8
Ewe's Milk	100%	100%	min 50%	100%	min 50% or 100%	100%	100%	min 50%
Raw/Pasteurized Milk	R	R	ewes' R cows' P	R	P	R	R	P
Dry Matter	min 44%	UL	min 44%	UL	min 44%, 48%	44%	min 48%	min 44%
Fat in Dry Matter	min 48%	UL	min 48%	UL	min 48%	48%	min 48%	min 48%
NaCl	max 2.5%	max 2.5%	max 2.5%	UL	max 2.5%	max 2%	max 2.5%	1.9%
Producing Area of Slovakia	middle	middle	middle	west	middle	east	middle	middle
Package	plastic foil	plastic foil	plastic foil	plastic foil	paper + aluminium foil	plastic container	plastic foil	plastic foil

UL—unlabeled.

2.2. Determination of Physicochemical Properties

Fat was determined according to ISO 3433:2008 [23], dry matter according to ISO 5534:2004 [24]. Fat in dry matter was determined by calculation. pH was determined by pH meter Orion Star A211 (Thermo scientific, Renfrew, UK). The analysis of all physicochemical parameters were replicated three times.

2.3. Microbiological Analysis

Ten grams of each sample from each cheese group were put into a sterile bag under aseptic conditions and homogenized in 90 mL of sterile peptone/saline solution (0.87%) for 2 min in a Stomacher bagmixer blender (Interscience, Saint-Nom-la-Bretèche, France). Decimal dilutions were prepared according to ISO 6887–5 (2010) [25].

Coliform bacteria (CB) were determined according to ISO 4832:2006 [26]. Yeasts and moulds (Y) were determined according to ISO 21527-1:2010 [27]. Presumptive lactobacilli (PLb) were determined by enumeration of colonies after anaerobic cultivation on de Man, Rogosa and Sharpe (MRS) agar (HiMedia, Maharashtra, India) for 72 h at the temperature of 37 °C. Presumptive lactococci (PL) were determined by enumeration of colonies after aerobic cultivation on M17 agar (HiMedia) for 72 h at the temperature of 30 °C. Enterococci (E) were determined by enumeration of colonies after aerobic cultivation on Bile Esculine Azide (BEA) agar (Biokar, Allonne, France) for 24 h at the temperature of 37 °C according to ISO 27205:2010 [28]. Microbial counts were performed in triplicate.

2.4. Analysis of Volatile Organic Compounds by Electronic Nose

The previously described electronic nose (e-nose) (Heracles II, Alpha M.O.S., Toulouse, France) method [22] was used for volatile organic compounds analysis. There was 2.5 g of sample incubated statically in a 20 mL vial in a thermostat block at the temperature of 50 °C for 15 min (Autosampler, Alpha

M.O.S.) and 5 mL volume of the headspace gaseous compounds was withdrawn using a headspace autosampler syringe and dispensed into the e-nose injector for each analysis. The identification of the compounds was performed by matching the measured peaks with *Kovats* retention indices with NIST library (The National Institute of Standards and Technology library) (>50%) by software Alpha Soft V14 (Alpha M.O.S.). Each sample was weighed and placed in three different vials, each one was analyzed once.

2.5. Analysis of Volatile Organic Compounds by Gas Chromatography Mass Spectrometry

The head-space solid-phase microextraction (HS–SPME) method previously described [20] was used for sample extraction in a modified version. The amount of 2.5 g of sample was incubated statically in a 20 mL vial in a thermostat block at the temperature of 50 °C for 30 min (CombiPal automated sample injector 120, CTC Analytics AG, Zwingen, Switzerland), with an SPME fibre (1 cm; 50/30 µm DVB/CAR/PDMS) (Supelco, Bellefonte, PA, USA) placed in the CombiPal for each analysis. The fibre was initially conditioned by heating in the SPME Fiber Cleaning and Conditioning Station (placed in the CombiPal) at the temperature of 270 °C for 1 h. SPME extracts were desorbed in the GC injector at the temperature of 250 °C for 1 min and the fibre was cleaned in SPME Fiber Cleaning and Conditioning Station at 230 °C for 10 min.

The relative content (expressed in percentage) of samples was determined by gas chromatography mass spectrometry (GC 7890B–MS 5977A) (Agilent Technologies Inc., Santa Clara, CA, USA) equipped with column DB–WAXms (30 m × 0.32 mm × 0.25 µm; Agilent Technologies Inc., Santa Clara, CA, USA) operating with a temperature program and MS conditions [20]. The identification of compounds was carried out by comparing mass spectra (over 80% match) with a commercial database NIST®2017, and Wiley library, retention times of reference standards (ethyl acetate, hexanoic acid and isoamyl alcohol) comparison of data on occurrence in cheese from Slovakia with literature [5,20,29]. The relative content of determined compounds was calculated by dividing individual peak area by the total area of all peaks. Peaks under 1% were not counted. Each sample was measured in triplicate.

2.6. Statistical Analysis

Compounds identified by e-nose with a discriminant D > 0.900 were selected as significant sensors, based on which the PCA (Principal Component Analysis) was made by Alpha Soft V14 (Alpha M.O.S.) software.

The STATGRAPHICS Centurion (© StatPoint Technologies, Inc., Warrenton, VA, USA) and GraphPad Prism 6.01 (GraphPad Software Incorporated, San Diego, CA, USA) were used for statistical physicochemical, microbiological and GC–MS analysis. The ANOVA method complemented by the Test of Tukey´s Multiple Comparison Test and unpair t-test with a value of $p < 0.05$ was applied.

3. Results

3.1. Physicochemical Properties of "Bryndza" Cheese

The results of the physicochemical parameters of "bryndza" cheese are shown in Table 2. The content of dry matter and fat in the dry matter are the parameters of bryndza cheese prescribed by the legislation [30]. The variability of "bryndza" cheese dry matter content ranged from 2.34% to 5.19% for farm dairies (B1, B2, B4, B6, B7) samples and from 2.37% to 5.52% for industrial dairies (B3, B5, B8) ones. Higher coefficients of variation were found in the parameter of fat in dry matter, which ranged from 1.75% to 7.98% in "bryndza" cheese from farm dairies and from 2.71% to 9.77% in samples from industrial dairies. Mixed "bryndza" cheese (B3, B8) samples had the lowest average fat content. The pH value of bryndza cheese as an indicator of ripening was more stable in "bryndza" cheese from industrial dairies (coefficient of variation ranged from 0.99% to 1.69%) than in "bryndza" cheese from farm dairies (coefficient of variation ranged from 1.22% to 3.82%).

Table 2. Physicochemical properties of bryndza cheese (average values, means ± SD).

	B1	**B2**	**B3**	**B4**	**B5**	**B6**	**B7**	**B8**
F (%)	24.8 ± 2.1	24.9 ± 1.6	19.4 ± 2.5	24.8 ± 1.3	24.3 ± 3.3	26.4 ± 0.4	26.3 ± 3.1	22.4 ± 0.5
DM (%)	47.7 ± 1.1	48.9 ± 1.4	42.5 ± 2.0	54.1 ± 2.8	49.3 ± 2.7	52.0 ± 1.4	53.1 ± 2.2	45.5 ± 1.1
FDM (%)	51.8 ± 3.4	50.9 ± 2.2	45.6 ± 4.5	45.8 ± 1.4	49.1 ± 3.9	50.8 ± 0.9	49.5 ± 4.0	49.4 ± 1.3
pH	5.24 ± 0.2	5.09 ± 0.1	5.34 ± 0.1	5.24 ± 0.1	5.14 ± 0.1	5.21 ± 0.1	5.26 ± 0.2	5.24 ± 0.1

F—fat, DM—dry matter, FDM—fat in dry matter.

3.2. Microbiological Quality

The counts of microorganisms in the analyzed "bryndza" cheese samples are in Table 3. The coliform bacteria in the "bryndza" cheese produced from pasteurized milk (producer B8) were < 1 log CFU/g. In contrast, the counts of coliform bacteria ranged from 3.46 to 6.78 log CFU/g in "bryndza" cheese produced from unpasteurized ewe's milk.

Table 3. Microbiological quality of bryndza cheeses (values of geomean) (log CFU/g).

	B1	**B2**	**B3**	**B4**	**B5**	**B6**	**B7**	**B8**
CB	3.77	3.46	6.01	6.78	3.72	6.22	5.27	<1
E	5.35	7.36	7.35	7.34	4.59	7.19	6.47	2.37
PL	8.36	9.15	8.98	9.03	8.98	8.52	8.59	7.17
PLb	7.73	8.98	8.69	8.76	8.48	8.35	8.32	6.56
Y	6.62	6.21	6.26	4.65	6.52	5.96	6.19	5.80
DG	5.75	5.20	5.11	4.33	5.93	5.49	6.06	5.24

CB—coliform bacteria, E—enterococci, PL—presumptive lactococci, PLb—presumptive lactobacilli, Y—yeasts, DG—*Dipodascus geotrichum*.

Enterococci in the "bryndza" cheese from pasteurized milk reached an average value of 2.37 log CFU/g in the monitored period. Higher counts of enterococci were found in "bryndza" cheese from unpasteurized milk in the case of coliform bacteria than in "bryndza" cheese from pasteurized milk. The counts of enterococci in these samples ranged from 4.59 log CFU/g ("bryndza" cheese sample B5 was made from pasteurized cow's milk and unpasteurized ewe's milk) to 7.36 log CFU/g.

Presumptive lactococci were dominated microbiota of "bryndza" cheese. The counts ≥ 7.17 log CFU/g were found in all samples. The counts of presumptive lactobacilli in bryndza cheese from pasteurized milk reached levels 6.56 log CFU/g and from unpasteurized milk ≥ 7.73 log CFU/g.

Yeast counts varied from 4.65 log CFU/g to 6.62 log CFU/g. The differences between the counts of yeasts in "bryndza" cheese from unpasteurized and pasteurized milk are not as significant as in the case of other groups of monitored microorganisms. Species *Dipodascus geotrichum* (before *Geotrichum candidum*) was dominated yeast.

3.3. Analysis of VOC

The VOC of different chemical natures and varied sensory descriptors were separated by e-nose, based on head-space gas chromatography with a flame-ionization detector. The twenty VOCs were identified as sensors (markers) with D > 0.900 (Table 4). The aroma profiles of "bryndza" cheese were dominated by esters (7), aldehydes (4), alcohols (4), free fatty acids (2), ketones (2), and one monoterpene.

Table 4. Volatile organic compounds in bryndza cheese determined by electronic nose (e-nose) with D > 0.900.

Volatile Organic Compounds		*Kovats'* **Retention Index DB-5 Column**	*Kovats'* **Retention Index DB-1701 Column**	**Sensory Descriptor** [1]
Ketones	2-butanone	594	690	butter, cheese, chemical, chocolate, ethereal, gaseous
	2,3-butanonedione	589	690	butter, caramelized, creamy, fruity, pineapple, spirit
Aldehyde	propanal	489	579	ethereal, plastic, pungent, solvent
	butanal	578	668	chocolate, green, malty, pungent
	heptanal	901	986	citrus, fatty, fruity, green, smoky
	2-methyl propanal	522	626	brunt, fruity, green, toasted, spicy, malty, pungent
	furfural	836	978	almond, bread, sweet
Esters	ethyl acetate	614	677	acidic, butter, caramelized, fruity, orange, pineapple, pungent, solvent, ethereal, sweet
	butyl acetate	813	879	banana, bitter, ethereal, green, strong, fruity, pear, pineapple, sweaty, sweet
	ethyl butanoate	800	865	acetone, banana, bubblegum, caramelized, fruity, pineapple, strawberry, sweet
	isoamyl acetate	876	945	banana, fresh, fruity, pear, sweet
	ethyl propanoate	710	766	acetone, fruity, solvent
	methyl 2-methyl butanoate	774	840	apple, chewing gum, fruity, solvent, spirit
Alcohols	2-methyl propanol	626	736	alcoholic, bitter, chemical, glue, leek, lecorice, solvent, winey
	2-propanol	500	602	alcoholic, ethereal
	n–butanol	664	779	cheese, fermented, fruity, medicinal
	1-hexanol	868	980	dry, floral, fruity, grassy, green, herbaceous, mild woody
Free Fatty Acids	propanoic acid	739	889	acidic, pungent, rancid, soy
	hexanoic acid	996	1186	cheese, fatty, goat, pungent, rancid, sweaty
Monoterpenes	limonene	1049	1073	citrus, fruity, minty, orange, peely

[1] Sensory descriptor from AroChemBase database, part of software Alpha Soft V14 (Alpha M.O.S.).

The dataset of e-nose was in the two-dimensional (2D) PCA plot. The first two principal components accounted for over 90% of total variance indicating that the first two PCs are sufficient enough to explain the total variance of the dataset. When the samples overlap or close to each other, it means they have a similar aroma as it is shown in Figure 1, 40 "bryndza" pieces of cheese from 8 producers divided into 5 groups according to the month of production in PC1 (76.25%). The samples from the May group were far away from the September group, which means that they have significantly different aromas and simultaneously, May and September groups of samples are surrounded by June, July, and August. In general, the PCA result suggested that the e-nose can properly characterize the samples of "bryndza" cheese.

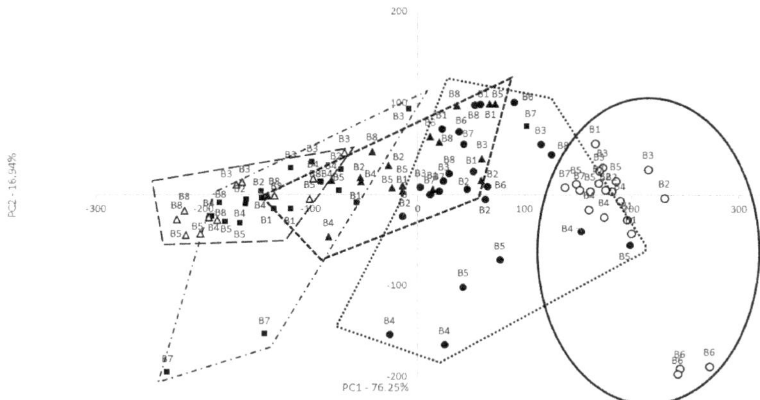

Figure 1. Projection of "bryndza" cheese from 8 producers (B1–8) onto the space defined by the first two principal components (PC1/PC2) based on the e-nose results. Sample groups according to the month: ○—May, ●—June, ▲—July, ■—August, △—September.

Based on the qualitative analysis of HS–SPME/GC–MS results, a total of 24 VOCs were identified from "bryndza" cheese produced by 8 producers including alcohols (7), free fatty acids (6), esters (4), aldehydes and ketones (4), one monoterpene and one oxime in this study. Data in Table 5 shown the average chemical composition of samples produced during 2019 (from May to September). The compounds with the highest relative contents in "bryndza" were isoamyl alcohol (2.86–10.6%), acetoin (6.16–20.7%), acetic acid (7.42–11.3%), butanoic acid (ND–5.88%), and methoxy-phenyl-oxime (3.74–8.54%).

Table 5. Average chemical composition (TIC% Area) [1] of the bryndza cheese from producers B1–8 produced from May to September 2019 obtained by gas chromatography (GC–MS) analysis.

Volatile Organic Compounds (TIC% Area)		B1	B2	B3	B4	B5	B6	B7	B8
	isoamyl alcohol [a,2]	5.20	4.64	4.07	3.22	10.6	4.34	2.86	4.56
	2,7-dimethyl-4,5-octandiol [a]	1.19	1.39	1.01	1.06	ND [3]	0.93	0.72	0.94
	2,3-butanediol	1.19	1.60	3.30	3.94	2.55	2.03	1.40	2.56
Alcohols	β-phenyl ethanol [a,b]	0.72	0.75	6.73	0.97	2.40	1.52	1.07	6.82
	dl-erythro-1-phenyl-1,2-propanediol [a]	1.78	0.64	0.71	0.60	ND	ND	ND	ND
	1-methoxy 2-propanol	ND	ND	ND	ND	1.59	ND	ND	ND
	2-butanol	ND	ND	ND	12.36	ND	ND	ND	ND
Aldehyde	Benzaldehyde [a]	2.04	0.79	0.55	0.48	ND	ND	0.62	ND
	acetoin	6.16	11.3	11.8	15.2	18.5	20.7	18.9	10.6
Ketones	2,3-butanedione [b]	ND	2.55	3.55	1.95	2.39	1.31	3.15	3.24
	2-butanone	ND	ND	ND	16.47	ND	ND	2.77	ND
	acetic acid	7.42	10.4	9.87	11.3	9.42	7.97	8.95	8.74
	butanoic acid [a]	4.33	4.02	2.00	1.80	2.45	3.01	5.88	ND
Free fatty	pentanoic acid [a]	1.15	1.50	1.00	ND	0.74	0.74	1.37	2.69
acids	hexanoic acid [a]	3.71	4.04	3.74	2.85	3.96	3.94	6.98	3.66
	octanoic acid	ND	1.76	1.97	1.37	4.18	3.55	4.23	1.52
	n-decanoic acid	ND	ND	ND	ND	2.51	ND	2.37	ND

Table 5. *Cont.*

Volatile Organic Compounds (TIC% Area)		B1	B2	B3	B4	B5	B6	B7	B8
Esters	ethyl acetate	3.10	5.26	6.10	3.63	3.63	8.07	ND	4.65
	2-phenetyl acetate [a,b]	0.93	1.37	8.07	0.85	ND	0.60	ND	3.34
	acetoin acetate [a]	ND	0.77	1.97	ND	1.14	0.90	ND	1.05
	isoamyl acetate	ND	ND	1.76	ND	ND	ND	ND	3.04
Monoterpenes	D-limonene [a,b]	1.05	0.76	0.50	ND	ND	ND	0.71	ND
Oxime	phenyl-methoxy-oxime	8.54	5.69	4.40	4.33	4.20	3.74	4.46	5.15

[1] Listed are the compounds that represented min. 1% in at least one bryndza cheese. [2] Letters in superscript indicates statistically significant difference ($p < 0.05$): [a]—among samples depending on type of milk; [b]—among samples depending on dairies. [3] ND—not detected.

The six compounds identified by e-nose with a D > 0.900 with ethyl acetate, isoamyl acetate, 2-butanone, hexanoic acid, D-limonene, and 2,3-butanedione were confirmed by GC-MS in this study. Statistical analysis of variance was used to differentiate cheese by evaluation scores such as type of milk, and dairies (farm or industrial) (Table 5). Particularly type of milk had a significant influence on the amounts of identified VOC. The relative chemical composition (TIC% Area) of 2,3-butanediol, 1-methoxy 2-propanol, 2-butanol, acetoin, 2-butanone, acetic acid, octanoic acid, *n*-decanoic acid, ethyl acetate, isoamyl acetate, and phenyl-methoxy-oxime were not significantly influenced by type of milk or dairies. Statistical analysis by the Test of Tukey's confirmed that TIC% Area of ten compounds was influenced by type of milk. The TIC% Area of β-phenyl ethanol, 2,3-butanedione, 2-phenetyl acetate, and D-limonene was significantly influenced by dairies ($p < 0.05$).

4. Discussion

The quality of "bryndza" cheese is affected by many factors, such as the composition of used ewe's and cow's milk, the seasonality of ewe's milk, or the microbiota of used starter culture [31–33]. "Bryndza" cheese must have a dry matter content at least 44 wt.% and fat in dry matter of ewe's "bryndza" cheese should be at least 48 wt.% and in mixed "bryndza" cheese at least 38 wt.% [30].

The reason for fat, fat in dry matter and dry matter content fluctuation in "bryndza" cheese from farm dairies is that the fat content in milk is not standardized in the traditional production. It is known that the fat and dry matter content of ewe's milk changes naturally during the season [34]. Industrial dairies also use cow's milk to "bryndza" cheese production, which explains lower fat content in the "bryndza" cheese from these producers (B3 and B8). Most "bryndza" cheese samples from the industrial producer (B3) had lower dry matter content than it is in legislative requirements. The changes in the composition of ewe's milk and in the produced lump cheese, as well during the season were also reflected in the sensory properties of "bryndza" cheese.

Coliform bacteria are considered as indicators of faecal contamination or poor hygienic conditions in obtaining and processing milk into cheese or low counts of competing LAB. The counts of LAB were high in the analyzed "bryndza" samples, therefore, the coliform bacteria counts indicate poor hygiene in milk processing into cheese. The exception is the counts of coliform bacteria in the "bryndza" cheese from farmers B1 and B2. The differences in the counts of coliform bacteria between "bryndza" cheese made from unpasteurized and pasteurized milk are in line with the results of other studies in this area [35]. Coliform bacteria in "bryndza" cheese have also been found by other authors [7,32,33].

Higher counts of enterococci were found in "bryndza" cheese samples made from unpasteurized ewe's milk on farms. Some authors [36] even found higher counts of enterococci (8.0 log CFU/g) in "bryndza" cheese made from raw ewe's milk in comparison with our results. In contrast, there are some results with very similar counts of enterococci (5.11–5.85 log CFU/g) in "bryndza" cheese made from raw and pasteurized milk in published studies [33]. The presence of enterococci in food may not always indicate faecal contamination, but rather, a violation of hygiene and sanitation principles. In the case of some food (cheese and fermented meat products), enterococci are added into them in a targeted way to improve organoleptic properties [37]. Enterococci and yeast in ewe's cheese made from raw milk contribute to the formation of acetic acid esters [38].

Various genera and species of LAB are present in the microbiota of raw ewe's milk. Their counts gradually increase during the production of lump cheese and "bryndza" cheese [7]. Starter culture, which is not so varied in genus and species, must be added into the milk after pasteurization to achieve lactic acidification. "Bryndza" cheese made from pasteurized milk may have lower counts of lactic acid bacteria, especially lactobacilli, as it is evidenced by our results and the works of other authors [32]. LAB break down lactose into lactic acid and various aroma compounds such as diacetyl, acetoin, acetaldehyde, or acetic acid [39]. However, they contribute very little to the lipolysis of milk fat. LAB cause the degradation of casein fractions into small peptides and free amino acids by the production of proteinases and peptidases. Amino acids can change into various alcohols, aldehydes, acids, esters, and sulfur compounds, which contribute to the specific aroma of the cheese [40].

Yeast is a natural microbiota of "bryndza" cheese and acts in the secondary stage of this cheese ripening. It is not surprising that the differences in yeast counts in "bryndza" cheese, made from raw and pasteurized milk, are minimal. *Dipodascus geotrichum* has a dominant position among yeasts in bryndza cheese. This type of yeast grows on the surface of the lump and breaks down lactates, milk fat (free fatty acids are released) and proteins (peptides and amino acids) [41,42]. Some strains of *Dipodascus geotrichum* are able to produce esters and various sulfur compounds, which contribute to the formation of a typical aroma and overall quality parameters of cheese [41].

LAB, enterococci, and yeasts *Galactomyces candidus* [6,7,32,43] play a key role in aroma development during cheese ripening. The VOCs are generated by the enzymatic degradation of amino acids in cheese, especially in cheese containing only LAB. The amino acid transamination is catalyzed by lactococci aminotransferases and this is the first step in the degradation of volatile and branched-chain amino acids, which are precursors of volatile organic compounds [44,45]. The resulting α-ketoacids are then degraded to aldehydes, alcohols, carboxylic acids, esters, methanethiol, and other sulfur compounds [46]. The seven compounds as 2-methyl propanol, ethyl acetate, ethyl butanoate, 2-butanone, isoamyl acetate, hexanoic acid, and butane-2,3-dione were identified in this study by e-nose and gas chromatography-olfactometry, as previously described in "bryndza" cheese from Slovakia [4,5,20,32]. Several of the identified compounds (ethyl acetate, 2-phenyl ethyl acetate, ethyl propanoate, 2-methyl-propanol, 2-phenyl ethanol, 2,3-butanediol, 3-hydroxy-2-butanone (acetoin), 3-methyl butanol, 2,3-butanedione, acetic, butanoic, pentanoic, and octanoic acids) are known to be components of different foreign ewe's cheese as the Oscypek, Canestrato Pugliese, Fiore Sardo, Torta del Casar, Terrincho, Roncal, Manchego, Pecorino Romano [47–49]. Other identified compounds, benzaldehyde [13], 2,7-dimethyl-4,5-octanediol [50] were previously described as secondary metabolites by LAB. Passerini et al. [51] confirmed that strains of *L. lactis* with the citP gene and the citM-G cluster produced a larger amount of VOCs than the strains without this genetic information. Likewise, the quality differences in milk and dairy products from different grazing areas have been previously reported [52,53].

The terpenes composition (limonene, myrcene, carvone) of milk and cheese are directly transferred from ingested botanical species and free fatty acids (acetic, butanoic, pentanoic, octanoic, decanoic, and dodecanoic acids) can also be effective to trace animal management and feeding systems [54]. Fatty acids were the most abundant VOCs in the barrelled ewe's cheese (intermediate product in the production of winter bryndza) from Slovakia [4] and in the raw ewes' milk cheese Torta del Casar [47,48] or Feta cheese [52]. The free fatty acids are also precursors of methyl ketones, alcohols, lactones, and esters, so they may play an important role in the global aroma development of cheese [48]. The identified 2-methyl propanal was previously described [55] as milk aromas. The other compounds, *n*-butanol and methoxy-phenyl-oxime were previously identified in cheese produced from pasteurization milk fermented with cultures mixed (*L. bulgaricus* and *Streptococcus thermophilus*) and with *Dregea sinensis* Hemsl. protease [56].

The e-nose technology in this study can detect the fingerprint of VOCs present in the headspace of the "bryndza" sample and "parenica" cheese previously described [22] by means of a gas chromatography principle. There are several studies in which an e-nose method containing 10 metal-oxide semiconductors for characterization of the volatile profile of French cheese types [57], Danish blue cheese [58], or Pecorino cheese were used [59]. These mentioned works used e-nose with sensors and it could not determine and identify concrete VOCs and therefore there was a need to confirm the results by GC methods.

5. Conclusions

The composition, microbiological quality, as well as the production of volatile organic compounds, changed during the ewe's milk season and "bryndza" cheese production by farm and industrial producers from Slovakia. The type of heat treatment of the milk, and the technology of the "bryndza" cheese production, also had an impact on the microbiological quality. There were coliform bacteria, enterococci, presumptive lactococci, presumptive lactobacilli, and yeasts detected. The key VOC were defined in "bryndza" cheese by e-nose and GC–MS. Based on the PCA result, the May samples had significantly different aromas, compared with the September samples.

Author Contributions: Conceptualization, J.Š. and M.Č.; methodology and laboratory analyses, J.Š., V.D., M.M., M.K. and M.Č.; data curation, V.D. and M.M.; writing—original draft preparation, J.Š., V.D., M.M., M.K. and M.Č.; writing—review and editing, J.Š., V.D. and M.Č. All authors have read and agreed to the published version of the manuscript.

Funding: This research was funded by the APVV–16–0244 grant "Qualitative factors affecting the production and consumption of milk and cheese".

Acknowledgments: This study was supported by Research Center AgroBioTech built in accordance with the project Building Research Center "AgroBioTech" ITMS 26220220180 and to Operational Program Research and Innovation: "Support of research activities in the ABT RC, 313011T465, co-financed by the European Regional Development Fund".

Conflicts of Interest: The authors declare no conflict of interest.

References

1. Zajác, P.; Martišová, P.; Čapla, J.; Čurlej, J.; Golian, J. Characteristics of textural and sensory properties of Oštiepok cheese. *Potr. Slovak J. Food Sci.* **2019**, *13*, 116–130. [CrossRef]
2. Šnirc, M.; Árvay, J.; Král, M.; Jančo, I.; Zajác, P.; Harangozo, Ľ.; Benešová, L. Content of mineral elements in the traditional Oštiepok cheese. *Biol. Trace Elem. Res.* **2020**, *196*, 639–645. [CrossRef]
3. Commission Regulation (Ec) No 676/2008 of 16 July 2008 registering certain names in the Register of protected designations of origin and protected geographical indications (Ail de la Drôme (PGI), Všestarská cibule (PDO), Slovenská bryndza (PGI), Ajo Morado de Las Pedroñeras (PGI), Gamoneu or Gamonedo (PDO), Alheira de Vinhais (PGI), Presunto de Vinhais Or Presunto Bísaro de Vinhais (PGI)), L189. *Off. J. Eur. Union* **2008**, *51*, 19–20.
4. Sádecká, J.; Šaková, N.; Pangallo, D.; Koreňová, J.; Kolek, E.; Puškárová, A.; Bučková, M.; Valík, L.; Kuchta, T. Microbial diversity and volatile odour-active compounds of barrelled ewes´ cheese as an intermediate product that determines the quality of winter bryndza cheese. *LWT Food Sci. Technol.* **2016**, *70*, 237–244. [CrossRef]
5. Šaková, N.; Sádecká, J.; Lejková, J.; Puškárová, A.; Koreňová, J.; Kolek, E.; Valík, Ľ.; Kuchta, T.; Pangallo, D. Characterization of May bryndza cheese from various regions in Slovakia based on microbiological, molecular and principal volatile odorants examination. *J. Food Nutr. Res. Slovak* **2015**, *54*, 239–251.
6. Kačániová, M.; Nagyová, Ľ.; Štefániková, J.; Felšöciová, S.; Godočiková, L.; Haščík, P.; Horská, E.; Kunová, S. Characterization of bryndza cheese from different regions of Slovakia based on microbiological quality. *Potr. Slovak J. Food Sci.* **2020**, *14*, 69–75. [CrossRef]
7. Pangallo, D.; Šaková, N.; Koreňová, J.; Puškárová, A.; Kraková, L.; Valík, L.; Kuchta, T. Microbial diversity and dynamics durnig the production of May bryndza cheese. *Int. J. Food Microbiol.* **2014**, *170*, 38–43. [CrossRef]

8. Ozturkoglu-Budak, S.; Wiebenga, A.; Bron, P.A.; de Vries, R.P. Protease and lipase activities of fungal and bacterial strains derived from an artisanal raw ewe's milk cheese. *Int. J. Food Microbiol.* **2016**, *237*, 17–27. [CrossRef]

9. Čaplová, Z.; Pangallo, D.; Kraková, L.; Puškárová, A.; Drahovská, H.; Bučková, M.; Koreňová, J.; Kuchta, T. Detection of genes prtP, pepN, pepX and bcaT involved in formation of aroma-active compounds in lactic acid bacteria from ewes´ cheese. *J. Food Nutr. Res. Slovak* **2018**, *57*, 195–200.

10. Yvon, M.; Rijnen, L. Cheese flavor formation by amino acid catabolism. *Int. Dairy J.* **2001**, *11*, 185–201. [CrossRef]

11. Dias, B.; Weimer, B. Conversion of methionine to thiols by Lactococci, Lactobacilli, and Brevibacteria. *App. Environ. Microbiol.* **1998**, *64*, 3320–3326. [CrossRef]

12. Dias, B.; Weimer, B. Purification and characterization of L-Methionine g-lyase from Brevibacterium linens BL2. *App. Environ. Microbiol.* **1998**, *64*, 3327–3331. [CrossRef]

13. Rijnen, L.; Delacroix-Buchet, A.; Demaizieres, D.; Le Quéré, J.L.; Gripon, J.C.; Yvon, M. Inactivation of lactococcal aromatic aminotransferase prevents the formation of floral aroma compounds from aromatic amino acids in semi-hard cheese. *Int. Dairy J.* **1999**, *9*, 877–885. [CrossRef]

14. Bourdat-Deschamps, M.; Le Bars, D.; Yvon, M.; Chapot-Chartier, M.P. Autolysis of Lactococcus lactis AM2 stimulates the formation of certain aroma compounds from amino acids in a cheese model. *Int. Dairy J.* **2004**, *14*, 791–800. [CrossRef]

15. Nierop-Groot, M.N.; de Bont, J.A.M. Conversion of phenylalanine to benzaldehyde initiated by an aminotransferase in Lactobacillus plantarum. *App. Environ. Microbiol.* **1998**, *64*, 3009–3013. [CrossRef]

16. Nierop-Groot, M.N.; de Bont, J.A.M. Involvement of manganese in conversion of phenylalanine to benzaldehyde by lactic acid bacteria. *App. Environ. Microbiol.* **1999**, *65*, 5590–5593. [CrossRef]

17. González-Martín, M.I.; Vivar-Quintana, A.M.; Revilla, I.; Salvador-Esteban, J. The determination of fatty acids in cheeses of variable composition (cow, ewe´s, and goat) by means of near infrared spectroscopy. *Microchem. J.* **2020**, *156*, 104854. [CrossRef]

18. Signorelli, F.; Contarini, G.; Annicchiarico, G.; Napolitano, F.; Orrú, L.; Catillo, G.; Haenlein, G.F.W.; Moioli, B. Breed differences in sheep milk fatty acid profiles: Opportunities for sustainable use of animal genetic resources. *Small Rumin. Res.* **2008**, *78*, 24–31. [CrossRef]

19. Erbay, Z.; Koca, N. Effects of using whey and maltodextrin in white cheese powder production on free fatty acid content, nonenzymatic browning and oxidation degree during storage. *Int. Dairy J.* **2019**, *96*, 1–9. [CrossRef]

20. Sádecká, J.; Kolek, E.; Pangallo, D.; Valík, L.; Kuchta, T. Principal volatile odorants and dynamics of their formation during the production of May Bryndza cheese. *Food Chem.* **2014**, *150*, 301–306. [CrossRef]

21. Boltar, I.; Čanžek Majhenič, A.; Jarni, K.; Jug, T.; Bavcon Krajl, M. Research of volatile compounds in cheese affected by different technological parameters. *J. Food Nutr. Res. Slovak* **2019**, *58*, 75–84.

22. Štefániková, J.; Nagyová, V.; Hynšt, M.; Vietoris, V.; Martišová, P.; Nagyová, Ľ. Application of electronic nose for determination of Slovak cheese authentication based on aroma profile. *Potr. Slovak J. Food Sci.* **2019**, *13*, 262–267. [CrossRef]

23. ISO. *Cheese—Determination of Fat Content—Van Gulik Method. International Organization for Standardization*; ISO 3433:2008; International Organization for Standardization: Geneva, Switzerland, 2008.

24. ISO. *Cheese and Processed Cheese Determination of the Total Solids Content (Reference Method)*; ISO 5534:2004; International Organization for Standardization: Geneva, Switzerland, 2004.

25. ISO. *Microbiology of Food and Animal Feeding Stuffs—Preparation of Test Samples, Initial Suspension and Decimal Dilutions for Microbiological Examination—Part 5: Specific rules for the Preparation of Milk and Milk Products*; ISO 6887-5:2010; International Organization for Standardization: Geneva, Switzerland, 2010.

26. ISO. *Microbiology of Food and Animal Feeding Stuffs—Horizontal Method for the Enumeration of Coliforms—Colony-Count Technique*; ISO 4832:2006; International Organization for Standardization: Geneva, Switzerland, 2006.

27. ISO. *Microbiology of Food and Animal Feeding Stuffs. Horizontal Method for the Enumeration of Yeasts and Moulds. Part 1: Colony Count Technique in Products with Water Activity Greater than 0.95*; ISO 21527-1:2010; International Organization for Standardization: Geneva, Switzerland, 2010.

28. ISO. *Fermented Milk Products—Bacterial Starter Cultures—Standard of Identity*; ISO 27205:2010; International IDF Standard: Brussels, Belgium, 2010.

29. Tomáška, M.; Čaplová, Z.; Sádecká, J.; Šoltys, K.; Kopuncová, M.; Budiš, J.; Drončovský, M.; Kolek, E.; Koreňová, J.; Kuchta, T. Microorganisms and volatile aroma-active compounds in "nite" and "vojky" cheeses. *J. Food Nutr. Res. Slovak* **2019**, *58*, 187–200.

30. Regulation no. 343/2016 of the Ministry of Agriculture and Rural Development of the Slovak Republic of 8 December 2016 on Certain Dairy Products. Available online: https://www.slov-lex.sk/pravne-predpisy/SK/ZZ/2016/343/ (accessed on 6 July 2020).

31. Planý, M.; Kuchta, T.; Šoltýs, K.; Semes, T.; Pangallo, D.; Siekel, P. Metagenomic analysis of Slovak bryndza cheese using nextgeneration 16SrDNA amplicon sequencing. *Nova Biotechnol. Chim.* **2016**, *15*, 23–34. [CrossRef]

32. Sádecká, J.; Čaplová, Z.; Tomáška, M.; Šoltys, K.; Kopuncová, M.; Budiš, J.; Drončovský, M.; Kolek, E.; Koreňová, J.; Kuchta, T. Microorganisms and volatile aroma-active compounds in bryndza cheese produced and marketed in Slovakia. *J. Food Nutr. Res. Slovak* **2019**, *58*, 382–392.

33. Semjon, B.; Reitznerová, A.; Poláková, Z.; Výrostková, J.; Maľová, J.; Koréneková, B.; Dudrikova, E.; Lovayová, V. The effect of traditional production methods on microbial, physico-chemical and sensory properties of Slevenska bryndza Pretected Geographical Indication cheese. *Int. J. Dairy Technol.* **2018**, *71*, 709–716. [CrossRef]

34. Dudrikova, E. *Hygienic and Technological Aspects of the Collection and Processing of Sheep´s Milk into the Mountain Conditions in Slovakia*, 1st ed.; University of Veterinary Medicine and Pharmacy: Košice, Slovakia, 2011; p. 121. (In Slovak)

35. Martin, N.H.; Trmčić, A.; Hsieh, T.H.; Boor, K.J.; Wiedmann, M. The evolving role of coliforms as indicators of unhygienic processing conditions in dairy foods. *Front. Microbiol.* **2016**, *7*, 1–8. [CrossRef]

36. Vrabec, M.; Lovayová, V.; Dudriková, K.; Gallo, J.; Dudriková, E. Antibiotic reistance and prevalence of Enterococcus sp. and Escherichia coli isolated from bryndza cheese. *Ital. J. Anim. Sci.* **2015**, *4*, 609–614.

37. Cocolin, L.; Foschino, R.; Comi, G.; Grazia Fortina, M. Description of the bacteriocins produced by two strains of Enterococcus faecium isolated from Italian goat milk. *Food Microbiol.* **2007**, *24*, 752–758. [CrossRef]

38. Liu, S.Q.; Holland, R.; Crow, V.L. Esters and their biosynthesis in fermented dairy products: A review. *Int. Dairy J.* **2004**, *14*, 923–945. [CrossRef]

39. Smit, G.; Smit, B.A.; Engels, W.J.M. Flavour formation by lactic acid bacteria and biochemical flavour profiling of cheese products. *FEMS Microb. Rev.* **2005**, *29*, 591–610. [CrossRef]

40. Van Kranenburg, R.; Kleerebezem, M.; van Hylckama Vlieg, J.E.T.; Ursing, B.M.; Boekhorst, J.; Smit, B.A.; Ayad, E.H.E.; Smit, G.; Siezen, R.J. Flavour formation from amino acids by lactic acid bacteria: Predictions from genome sequence analysis. *Int. Dairy J.* **2002**, *12*, 111–121. [CrossRef]

41. Koňuchová, M.; Liptáková, D.; Šípková, A.; Valík, Ľ. Role of Geotrichum candidum in Dairy Industry. *Chem. Listy* **2016**, *110*, 491–497.

42. Jaster, H.; Judacewski, P.; Ribeiro, J.C.B.; Zielinski, A.A.F.; Demiate, I.M.; Los, P.R.; Alberti, A.; Nogueira, A. Quality assessment of the manufacture of new ripened soft cheese by Geotrichum candidum: Physico-chemical and technological properties. *Food Sci. Technol.* **2018**, *39*, 50–58. [CrossRef]

43. Kačániová, M.; Kunová, S.; Štefániková, J.; Felšöciová, S.; Godočíková, L.; Horská, E.; Nagyová, Ľ.; Haščík, P.; Terentjeva, M. Microbiota of the traditional Slovak sheep cheese "Bryndza". *J. Microbiol. Biotechnol. Food Sci.* **2019**, *9*, 482–486. [CrossRef]

44. Yvon, M.; Thirouin, S.; Rijnen, L.; Fromentier, D.; Gripon, J.C. An aminotransferase from Lactococcus lactis initiates conversion of amino acids to cheese flavor compounds. *App. Environ. Microbiol.* **1997**, *63*, 414–419. [CrossRef] [PubMed]

45. Tanous, C.; Gori, A.; Rijnen, L.; Chambellon, E.; Yvon, M. Pathways for α-ketoglutarate formation by Lactococcus lactis and their in amino acid catabolism. *Int. Dairy J.* **2005**, *15*, 759–770. [CrossRef]

46. Savijoki, K.; Ingmer, H.; Varmanen, P. Proteolytic systems of lactic acid bacteria. *App. Microbiol. Biotechnol.* **2006**, *71*, 394–406. [CrossRef]

47. Delgado, F.J.; González-Crespo, J.; Cava, R.; García-Parra, J.; Ramírez, R. Characterization of the volatile profile of a Spanish ewe raw milk soft cheese P.D.O. Torta del Casar during ripening by SPME-GC-MS. *Food Chem.* **2010**, *118*, 182–189. [CrossRef]

48. Delgado-Martínez, F.J.; Carrapiso, A.I.; Contador, R.; Rosario Ramírez, M. Volatile compounds and sensory changes after high pressure processing of mature "Torta del Casar" (raw ewe´s milk cheese) during refrigerated storage. *Innov. Food Sci. Emerg. Technol.* **2019**, *52*, 34–41. [CrossRef]

49. Majcher, M.A.; Goderska, K.; Pikul, J.; Jeleń, H.H. Changes in volatile, sensory and microbial profiles during preparation of smoked ewe cheese. *J. Sci. Food Agric.* **2011**, *91*, 1416–1423. [CrossRef]

50. Nájera-Domínguez, C.; Gutiérrz-Méndez, N.; Nevárez-Moorillon, G.; Caro-Canales, I. Comparison of volatile compounds produced by wild Lactococcus lactis in miniature Chihuahua-type cheeses. *Dairy Sci. Technol.* **2015**, *94*, 499–516. [CrossRef]

51. Passerini, D.; Laroute, V.; Coddeville, M.; Le Bourgeois, P.; Loubiere, P.; Ritzenthaler, P.; Cocaign-Bousquet, M.; Daveran-Mingot, M.L. New insights into Lactococcus lactis diacetyl- and acetoin producing strains isolated from diverse origins. *Int. J. Food Microbiol.* **2013**, *160*, 329–336. [CrossRef]

52. Bozoudi, D.; Kondyli, E.; Claps, S.; Hatzikamari, M.; Michaelidou, A.; Biliaderis, C.G.; Litopoulou-Tzanetaki, E. Compositional characteristics and volatile organic compounds of traditional PDO feta cheese made in two different mountainous areas of Greece. *Int. J. Dairy Technol.* **2018**, *71*, 673–682. [CrossRef]

53. Iussig, G.; Renna, M.; Gorlier, A.; Lonati, M.; Lussiana, C.; Battaglini, L.M.; Lombardi, G. Browsing ratio, species intake, and milk fatty acid composition of goats foraging on alpine open grassland and grazable forestland. *Small Rumin. Res.* **2015**, *132*, 12–24. [CrossRef]

54. Moran, L.; Aldezabal, A.; Aldai, N.; Barron, L.J.R. Terpenoid traceability of commercial sheep cheeses produced in mountain and valley farms: From pasture to mature cheeses. *Food Res. Int.* **2019**, *126*, 108669. [CrossRef]

55. Cadwallader, K. Instrumental measurement of milk flavour and colour. In *Improving the Safety and Quality of Milk*, 1st ed.; Griffiths, M.W., Ed.; Woodhead Publishing: New York, NY, USA, 2010; Volume 2, pp. 181–206.

56. Wang, H.; Wang, Y.; Huang, A. Influence of Dregea sinensis Hemsl. protease on the quality of mozzarella cheese from buffalo milk. *Emir. J. Food Agric.* **2017**, *29*, 539–546. [CrossRef]

57. Ghasemi-Varnamkhasti, M.; Mohammad-Razdari, A.; Yoosefian, S.H.; Izadi, Z.; Siadat, M. Aging discrimination of French cheese types based on the optimization of an electronic nose using multivariate computational approaches combined with response surface method (RSM). *LWT* **2019**, *111*, 85–98. [CrossRef]

58. Trihaas, J.; Vognsen, L.; Nielsen, P.V. Electronic nose: New tool in modelling the ripening of Danish blue cheese. *Int. Dairy J.* **2005**, *15*, 679–691. [CrossRef]

59. Cevoli, C.; Cerretani, L.; Gori, A.; Caboni, M.F.; Gallina Toschi, T.; Fabbri, A. Classification of Pecorino cheeses using electronic nose combined with artificial neural network and comparison with GC–MS analysis of volatile compounds. *Food Chem.* **2011**, *129*, 1315–1319. [CrossRef]

© 2020 by the authors. Licensee MDPI, Basel, Switzerland. This article is an open access article distributed under the terms and conditions of the Creative Commons Attribution (CC BY) license (http://creativecommons.org/licenses/by/4.0/).

Article

Characterization of Whey-Based Fermented Beverages Supplemented with Hydrolyzed Collagen: Antioxidant Activity and Bioavailability

Arely León-López [1], Xóchitl Alejandra Pérez-Marroquín [1], Gieraldin Campos-Lozada [1], Rafael G. Campos-Montiel [1] and Gabriel Aguirre-Álvarez [1,2,*]

[1]   Instituto de Ciencias Agropecuarias, Universidad Autónoma del Estado de Hidalgo, Avenida Universidad Kilometro 1, Tulancingo C.P. 43600, Hidalgo, Mexico; arlely@hotmail.com (A.L.-L.); ale.marr28@gmail.com (X.A.P.-M.); gieraldin.campos@gmail.com (G.C.-L.); rcampos@uaeh.edu.mx (R.G.C.-M.)

[2]   Uni-Collagen S.A. de C.V., Arnulfo González No. 203, El Paraíso, Tulancingo C.P. 43684, Hidalgo, Mexico

*   Correspondence: aguirre@uaeh.edu.mx; Tel.: +52-775-145-9265

Received: 24 July 2020; Accepted: 10 August 2020; Published: 12 August 2020

Abstract: In this study, the preparation of a milk whey-based beverage with the addition of different concentrations of hydrolyzed collagen (0.3%, 0.5%, 0.75%, and 1%) was carried out. The control was considered at a concentration of 0%. Physicochemical properties, viscosity, antioxidant activity, and microbiological parameters were evaluated. The 1% collagen treatment showed the highest protein content (9.75 ± 0.20 g/L), as well as radical inhibition for ATBS (48.30%) and DPPH (30.06%). There were no significant differences ($p \geq 0.05$) in the fat and lactose parameters. However, the pH in the control treatment was lower compared to beverages treated with hydrolyzed collagen. Fourier transform-infrared spectroscopy showed spectra characteristic of lactose and collagen amides. The viscosity increased significantly as the concentration of hydrolyzed collagen increased. The addition of hydrolyzed collagen increased the bioavailability, nutritional value, and the antioxidant activity of the beverage. Hydrolyzed collagen acted as an antimicrobial agent, as there was no presence of microorganism pathogens observed in the treated beverages.

Keywords: milk whey; hydrolyzed collagen; antioxidant activity; bioavailability

1. Introduction

The food industry has undergone constant changes due to high demands for food and the requirement of satisfying the nutritional needs of consumers. Functional foods are the main source of development and innovation in the food industry. These types of foods have been recognized as having physiological benefits beyond those of basic nutrition. Functional foods can be classified as whole, fortified, enriched, or enhanced foods or food compounds that have health benefits for the human body [1–3]. There is a wide range of functional foods, including baby foods, baked goods, cereals, dairy products, confectionery products, meat products, and beverages [4]. Functional beverages can be supplemented or enriched with functional ingredients, such as vitamins, minerals, bioactive peptides, probiotics/prebiotics, etc. [5]. Over 40% of functional foods are dairy products, and fermented beverages containing milk whey are the principal functional dairy beverage [6]. Milk whey is a translucent greenish yellow liquid that is produced during the coagulation of casein in cheese production. There are 2 types of whey: sweet whey (pH 5.9 to 6.5), which is obtained from rennet coagulation in hard and soft cheeses, and acid whey (pH 4.4 to 4.8), which is produced by acid-coagulated fresh milk [7]. The processing of 1 kg of cheese produces approximately 9 L of whey [8]. This whey is then discarded without treatment to public sewage systems, generating a critical pollution problem. Approximately only 50% of the whey produced globally is used to formulate products. The remainder is treated as waste. Whey has traditionally been dumped into surface water or fed to livestock. However, current environmental regulations and levies are forcing cheese makers to treat whey before disposal, and because of factory centralization, the cost of transporting whole whey for feed use has become prohibitive. Therefore, whey has become a liability and a great amount of research has been focused on converting this liability into an asset [9]. Fortunately, milk whey has good antimicrobial, antioxidant, and antiviral properties. The biological activity of whey is related to the composition and sequence of the amino acids obtained by lactic acid bacteria (LAB) during fermentation [10]. Whey possesses nutrients such as calcium, phosphorus, magnesium, and the vitamins riboflavin and thiamine; represents about 85–95% of milk volume; and retains up to 55% of its nutrients after processing [11,12]. The proteins present in milk whey are α-lacto albumin (13%), β-lactoglobulin (58%), immunoglobulins, and a low concentration of serum albumin [13,14]. The preparation of fermented beverages containing milk whey has the disadvantage of low solids and casein content, giving a watery consistency to the final products. However, the addition of fresh milk, condensed milk, milk powder, or some other additives to improve the textural and nutritional characteristics of the final products is common [15]. The first step to obtain hydrolyzed collagen (HC) is the denaturation of the native collagen identified for the separation of 3 α-chains. After that, the proteolytic enzyme action of alcalase, papain, and other enzymes is used to initiate hydrolysis. HC is a group of peptides with low molecular weight, around 6 KDa, that can be extracted from different sources (e.g., bovine, porcine, fish, ovine) and through different methods (alkaline methods using NaOH or $NaHCO_3$, and acidic methods using acetic acid) [16]. Due to its low molecular weight, HC presents high biocompatibility, very easy degradation, and weak or no allergenicity [17]. In addition, HC is used as a functional ingredient within the food industry because of its antioxidant and antimicrobial activities. It helps to increase water holding, improving chemical and physical properties without modifying the sensorial properties in beverages and dairy products [18]. The objective of this research is to develop a beverage using milk whey and HC as functional ingredients to improve the nutritional, physicochemical properties, with antioxidant activity and bioavailability.

2. Materials and Methods

2.1. Milk Whey Characterization

The milk whey was donated by PROUNILAC Co. (Tulancingo, Hidalgo, Mexico). It was obtained as a byproduct from Oaxaca cheese processing by the coagulation of milk with organic acids. After slow pasteurization at 60 °C for 30 min, it was characterized for protein content [19], pH value [20], fat [21], and lactose [22].

2.2. Hydrolyzed Collagen

HC was obtained according to a previously determined methodology [23]. Native collagen was suspended in 1 M $NaCO_3$ at a ratio of 1:4 (*w/v*). The pH was adjusted to 8 ± 0.2. The hydrolysis was treated under enzymatic conditions for 2 h at 70 °C. The molecular weight was reported to be around 5.62 KDa with 0 cP of viscosity.

2.3. Preparation of Beverage

First, 100 mL of functional beverage was prepared with milk whey and HC according to the methodology of Tirado-Armesto [24], with some modifications. Then, 10% (*w/v*) free lactose powder milk and 1% (*w/v*) sucrose was added to the milk whey and homogenized for 10 min at 3200 rpm using an Ultra-Turrax T-25 Digital (IKA; Wilmington, NC, USA) dispersing instrument. Later, milk whey was fermented by LAB: *Lactobacillus rhamnosus*, *Lactobacillus bulgaricus*, *Lactobacillus delbrueckii*, and *Streptococcus thermophilus* (0.001% *w/v*) (DANISCO, Paris, France) for 3 h at 37 ± 2 °C. The pH was tested during the fermentation process, until 6.0 ± 0.2 was reached. The fermentation process was stopped by decreasing the temperature to 4 °C. Different HC supplementation was added as follows: 0% (control), 0.3%, 0.5%, 0.75%, and 1% (*w/v*). The beverage was pasteurized at 60 °C for 30 min. Samples were stored at 4 °C for further analysis.

2.4. Protein Determination

The Bradford assay was used to determine the protein concentration [19]: 100 µL of the sample were mixed with 5 mL of Bradford reagent (Thermo Fisher Scientific; MA, USA) in a vortex for 2 min. The sample was stored in darkness for 5 min. The absorbance of the sample was read at 595 nm in a spectrophotometer (Jenway Genova, Model 6705, Bibby Scientific; Staffordshire, UK). A calibration curve of serum albumin at different concentrations was used.

2.5. Fat Determination

According to the Association of Official Analytical Chemists (AOAC) method [21], 11 mL of sample and 1 mL of isoamyl alcohol (Merck Millipore; Bedford, MA, USA) were added in a Gerber butyrometer. Then, sulfuric acid was added to cover the sample with gently shaking in a water bath at 65 °C for 10 min. The sample was centrifuged in a Gerber Centrifuge (Zelian, model Galatea 24; Buenos Aires, Argentina) at 12,000 rpm for 3 min followed for another 5 min in a water bath to complete the separation of fat. Readings were taken and expressed in mL of fat.

2.6. pH Value

The pH measurements were made in triplicate following the methodology described by AOAC [20] in a previously calibrated potentiometer (HANNA Instruments, model HI 2210; Limena, Italy).

2.7. Lactose Determination

The reaction process of carbohydrates by the anthrone method is an extremely sensitive and specific reaction that can be applied to all carbohydrates, whether or not they are reducers [25]. First, 1 mL of the sample was mixed with 2 mL of anthrone reagent (Sigma-Aldrich; St. Louis, MI, USA) into a water bath at 70 °C for 10 min. The sample was cooled to room temperature and a reading was taken using a spectrophotometer (Jenway Genova, Model 6705, Bibby Scientific; Staffordshire, UK) at a wavelength of 625 nm [19]. The equipment was calibrated with lactose solutions prepared at different concentrations.

2.8. Ash Determination

Following the AOAC [26] methodology, 3 g of sample were weighed into a porcelain crucible that was previously subjected to constant weight. The crucible was placed into a muffle furnace (BIOBASE, 1200c; Qingdao, China) at 550 °C for 5 h until complete calcination occurred. The crucible was cooled to room temperature, followed by the determination of the ash content by using Equation (1):

$$\% \text{ Ash} = \frac{(C - A)}{B} \times 100, \tag{1}$$

where A is the crucible weight (g), B is the weight of the sample (g), and C is the weight of the crucible with ash (g).

2.9. Hydroxyproline Content

First, 4 g of the sample was mixed with 3.5 M sulfuric acid (30 mL) and incubated in an oven at 105 °C for 12 h. The volume was adjusted to 500 mL with distilled water followed by filtration. Then, 1 mL of oxidant solution (0.006 M chloramine T in 0.8 M citrate buffer, pH 6.0, Sigma-Aldrich; St. Louis, MI, USA) and 2 mL of the filtered sample were mixed in a reaction tube. The volume was adjusted to 100 mL and stirred for 30 min at 25 °C. Then, 2 mL of color reagent (10 g of dimethylamine benzaldehyde in 35 mL of 65% perchloric acid; Sigma-Aldrich; St. Louis, MI, USA) were added and stirred for 15 min at 60 °C [27]. The sample was read in a spectrophotometer (Jenway Genova, Bibby Scientific; Staffordshire, UK) at 558 nm. Equation (2) was used to calculated the hydroxyproline concentration:

$$\% \text{ Hydorxyproline} = \frac{(\text{Hydroxyproline concentration from the standard curve})(2.5)}{(\text{sample weight })(\text{volume in mL to adjust to 100 mL})}. \tag{2}$$

2.10. Viscosity Analysis

Before the viscosity measurements, the samples were conditioned at 4 °C. A viscosimeter (Brookfield RTV; Middleboro, MA, USA) was used to determinate the sample viscosity by using spindle number 5 and a speed of 50 rpm. Results were expressed in centipoise (cP).

2.11. Fourier Transform-Infrared Spectroscopy

Fourier transform-infrared spectroscopy (FTIR) (Perkin Elmer; Boston, MA, USA) equipment was used to obtain absorption spectra within the range of 380 to 4000 cm^{-1} wavelengths at room temperature. Samples were placed in intimate contact with the diamond crystal by applying a loading pressure. Four scans with an average of 4 cm^{-1} resolution were represented in each sample. Automatic signals were collected in 3620 scans with a resolution of 1 cm^{-1}. The spectrum of an empty cell was used as a background. Results were analyzed with Spectrum^{TM} 10 software (Perkin Elmer, Boston, MA, USA).

2.12. Differential Scanning Calorimetry

Differential scanning calorimetry equipment series Q 2000 with an intracooler RCS90 (TA Instruments; New Castle, Delaware. USA) was calibrated with indium (T_m, onset = 156.68 °C, ΔH = 28.45 J/g). Then, 1.5 ± 0.1 mg of the sample with 0% db water content was packed and hermetically sealed in a 50-mL stainless steel pan. An empty pan was used as a reference. Both heating and cooling scan rates were performed at 10 °C/min. Two heating scans were performed from 25 to 120 °C. TA 2000 analysis software (TA Instruments; New Castle, DE, USA) based on the endothermic changes registered in the thermogram was used to determine the melting temperature (T_m) and enthalpy (ΔH).

2.13. Antioxidant Activity

The ABTS (2,2′-Azino-bis(3-ethylbenzothiazoline-6-sulfonic acid)) radical solution was prepared according to Re et al.'s method [28]. First, 2.45 mM potassium persulfate were mixed with 7 mM of ABTS (Sigma-Aldrich; St. Louis, MI, USA). The mixture was stirred at room temperature for 16 h in the darkness. The ABTS solution was stabilized to 0.70 ± 0.02 at 734 nm using ethanol. Then, 2 mL of the sample and 1 mL of stabilized radical ABTS solution were mixed, and after 6 min, the sample was read at 734 nm in a spectrophotometer (Jenway Genova, Model 6705, Bibby Scientific; Staffordshire, UK).

A similar methodology was applied to samples for DPPH (2,2-diphenyl-1-picrylhydrazyl) antioxidant activity [29]. In this case, 0.5 mL of the sample was mixed with 2.5 mL of 6.1×10^{-5} M of DPPH radical inhibitor (Sigma-Aldrich; St. Louis, MI, USA). The mixture was stored in darkness for 30 min and then read in a spectrophotometer (Jenway Genova, Model 6705, Bibby Scientific; Staffordshire, UK) at 515 nm. The antioxidant activity of both ABTS and DPPH radical scavengers were calculated according to Equation (3):

$$\% \text{ Inhibition} = \frac{\text{Ia} - \text{Fa}}{\text{Ia}} \times 100, \tag{3}$$

where Ia is the initial absorbance and Fa is the final absorbance.

2.14. Microbiological Analysis

Determination of *Salmonella* and *E. coli* were followed according to Mexican legislation (NOM-210-SSA-2014) as an indicator of microorganisms of contamination, using Salmonella

Shigella Agar (Becton Dickinson; Heidelberg, Germany) and Eosin Methylene Blue Agar (Becton Dickinson; Heidelberg, Germany), respectively. For the determination of molds and yeasts in food (Sabouraud Dextrose Agar, Becton Dickinson; Heidelberg, Germany), the methodology of NOM-11-SSA-1994 was followed. In addition, NOM-092-SSA-1994 was followed for the aerobic bacteria count (aerobic mesophiles) by using Standard Methods Agar (DIBICO; Cuautitlan Izcalli, Mexico). The samples were incubated in petri dishes (90 × 15 mm) for 24 h in an oven (FELISA model FE-131; Zapopan, Jalisco. Mexico) at 37 ± 2 °C. The results were expressed as the presence or absence of microorganisms that are indicative of an efficient pasteurization process.

2.15. In Vitro Bioavailability of Hydrolyzed Collagen in Beverages

The gastric simulation methodology of Bilek and Bayram [30] was used with some modifications. First, 50 mL of the beverage was mixed with 6.5 mg of pepsin (15,750 units; Sigma-Aldrich; Munich, Germany). The pH of the sample-pepsin solution was adjusted to 2.0 with 1 M HCl and incubated at 37 °C for 2 h. Subsequently, 10 mL were taken and transferred to a dialysis membrane (2 KDa) and mixed with 2.5 mL of a bile acid-pancreatin (Sigma-Aldrich; Munich, Germany). The pH was adjusted to 7.5 with 0.5 M $NaHCO_3$. Along with this experiment, a buffer solution was prepared in a centrifuge tube with 10 mL of deionized water, the same amount of 0.5 M $NaHCO_3$, and adjusted to pH 5.0 with 1 M HCl. The dialysis membrane with the sample-pepsin solution was placed inside the centrifuge tube with a buffer and incubated at 37 °C for 2 h. The sample inside the membrane was subjected to hydroxyproline analysis to determine the amount of collagen present after gastric simulation. The percentage of bioavailability was calculated by using Equation (4):

$$\% \text{ Bioavailability} = \frac{\text{Collagen in dialysate (g)}}{\text{Collagen in sample}} \times 100. \tag{4}$$

2.16. Statistical Analysis

A randomized design experiment was applied. Three replicates were considered in this experiment. Analysis of variance (ANOVA) and Tukey tests ($p \leq 0.05$) were also used. Data were processed with SPSS software v25 (SPSS Inc.; Chicago, IL, USA).

3. Results and Discussion

3.1. Acid Milk Whey Chemical Characterization

The characteristics of the raw material were as follows: pH of 4.4 ± 0.2, 0.4 ± 0.4 g/L of total solids, 2.4 ± 0.4 g/L of total proteins, 29.3 ± 0.1 g/L of lactose, and 0.4 ± 0.1 g/L of fat. These results appeared above the average general composition of whey in terms of protein and lactose concentration [31]. Whey from semi-fat quark cheese production showed a higher solids concentration and pH, but a lower protein presence [32]. Whey from Ras cheese production resulted in higher fat and solids concentrations, but lower protein and lactose concentrations [33]. The composition of the milk whey depends on the source of the milk, type of cheese, and the cheese processing method [34–37]. The variability in the chemical composition might be due to the fact that the cheese production is a non-standardized process, so depending on the production methodology, the final characteristics of the whey will be affected [38,39].

3.2. Physicochemical Characterization of The Functional Beverage

Table 1 shows the functional beverage characterization. There was no significant difference ($p \leq 0.05$) between the resulting pH values and the treatments with HC (~7.0). However, the control treatment (0%) showed a lower pH value (5.1) compared to the others. There was a higher resulting pH in samples with HC because the hydrolysis of collagen was carried out in an alkaline medium ($NaHCO_3$). Beverages based on milk whey fermentation present some advantages: a decrease in lactose content, partial hydrolysis of whey protein, increase in the production of lactic acid, and production of aroma compounds that improve sensory characteristics [14,40]. However, there is a disadvantage in its low total solids content when compared to milk. Therefore, the addition of HC, sucrose, and milk powder could fix this problem. The acid lactic bacteria used for fermentation in this experiment (*L. rhamnosus*, *L. bulgaricus*, *L. delbrueckii*, and *S. thermophilus*) belonged to a group of lactose consuming microorganisms. These types of bacteria can improve the odor and flavor in fermented beverages [41]. The beverage protein concentration was 9.75 ± 0.20 g/L in the treatment with the highest collagen concentration (1%). It presented a significant difference ($p \leq 0.05$) compared to the treatments with the lowest collagen concentration and the control. These results suggest that the addition of HC could increase the nutritional value of the beverage. Hailu et al. [42] conducted a study based on whey beverages that presented high nutritional value and an ideal source of energy and nutrients. The ash content in the functional beverage provided an estimate of the minerals present in the sample. Minerals were found in ash as oxides, sulfates, phosphates, nitrates, chlorides, and other halides. The predominant minerals in milk whey are sodium and potassium [13,43]. The fat content in the functional beverage did not show a significant difference ($p \geq 0.05$) between the treatments; the values were between 0.20 ± 0.10 and 0.23 ± 0.06 g/L. These results are similar to those reported in previous works [44,45] based on the preparation of beverages with fermented milk whey and low fat content. The same behavior was observed for lactose parameters with no significant differences ($p \geq 0.05$) no matter what the amount of HC is.

Table 1. Functional beverage physicochemical characterization.

Hydrolyzed Collagen Concentration (%)	pH	Ash (%)	Fat (g/L)	Lactose (g/L)	Protein (g/L)
0	5.10 ± 0.03 [b]	0.82 ± 0.11 [c]	0.20 ± 0.10 [a]	27.09 ± 10.63 [a]	9.13 ± 0.09 [c]
0.3	7.07 ± 0.01 [a]	0.83 ± 0.11 [bc]	0.23 ± 0.06 [a]	18.65 ± 8.60 [a]	9.35 ± 0.10 [b]
0.5	7.03 ± 0.03 [a]	1.14 ± 0.02 [ab]	0.27 ± 0.06 [a]	20.10 ± 3.34 [a]	9.45 ± 0.07 [b]
0.7	7.36 ± 0.03 [a]	1.03 ± 0.01 [a]	0.27 ± 0.06 [a]	15.12 ± 3.62 [a]	9.48 ± 0.11 [b]
1.0	7.39 ± 0.03 [a]	1.19 ± 0.06 [a]	0.23 ± 0.06 [a]	28.26 ± 6.31 [a]	9.75 ± 0.20 [a]

Values are means $\pm$ SD based on 3 observations. Means with different lowercase letters differ ($p \leq 0.05$).

3.3. Viscosity Determination of the Whey Beverage

Figure 1 showed significant difference ($p \leq 0.05$) in viscosity between the highest concentration of HC addition (1%) and the lowest HC addition (0.3%). Viscosity increased from 395.55 ± 27.75 Cp to 537.7 ± 15.39 Cp, respectively. In dairy beverages, viscosity is an important characteristic for the consumer's acceptance and it depends on the solids content; type and concentration of additives; and fermentation conditions like time, temperature, and the LAB used [46]. The higher the HC concentration, the higher the viscosity observed in the beverages. HC is used in food to improve the stability because it is able to reduce the surface tension at the liquid interface by increasing the viscosity of the aqueous phase. In addition, it has been reported that HC presented greater water retention capacity, reducing syneresis and sedimentation in the beverage [47–49]. Previous works also presented a change in the viscosity in whey beverages by incorporating some other components, such as açaí pulp, HC, and sweet potato flour [49,50]. Additionally, the addition of soy protein in fermented whey beverages raised the viscosity [51]. Cassava starch is also a common option to obtain the desired viscosity and stability characteristics to improve the texture and sensory properties of whey beverages [52]. Dairy products are complex in composition and structures, showing differences in viscosity. The addition of some natural ingredients can help to achieve physicochemical properties like viscosity.

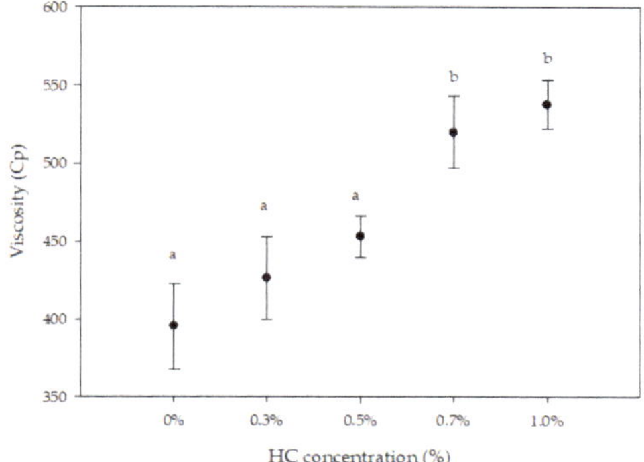

Figure 1. Beverage viscosity with different concentrations of hydrolyzed collagen (HC). Values are means ± SD based on 3 observations. Means with different lowercase letters differ ($p \leq 0.05$).

3.4. Fourier Transform-Infrared Spectroscopy

When using the FTIR technique, the samples do not require pretreatment or the use of any organic solvents, which reduces the environmental damage caused by toxic waste. The FTIR technique is based on natural vibrational frequencies of the chemical bonds present in the molecules [53]. As illustrated in Figure 2, the peaks between 1644 and 1649 cm^{-1} are representative of the fatty acids present in dairy products, and 1080 cm^{-1} is representative of lactose [54,55]. The characteristic peaks for collagen were observed at 3295 cm^{-1}, which is related to the tension vibrations of the NH group (Amide A), and 2946 cm^{-1}, which is related to the asymmetric stretching of the CH$_2$ groups (Amide B). Additionally, the peak at 1641 cm^{-1} is representative of Amide I, which is correlated with the α-helix chains and is used specifically to analyze the secondary structure of collagen. Amide II (1548 cm^{-1}) and Amide III (1248 cm^{-1}) were mainly associated with intermolecular interactions, representing the stretching vibrations of the CN group and the deformation of the NH group of the amide bonds, respectively. These peaks were observed in a previous work where HC was extracted [23]. The increase in the amplitude in the region of 3000 to 3500 cm^{-1} was also attributed to hydrogen bonds, due to the presence of the molecular chains interacting by inter- and intra-molecular hydrogen bonds [56]. The peak around 1000 cm^{-1} can be related to the presence of lactose and OH groups in the amino acid hydroxyproline, which is the main amino acid present in HC. The increment in amplitude of this peak is due to increased CH concentrations in the beverage [23,54]. Previous works also found Amide I and Amide II in the same region and these peaks were related to the amount of the protein present in the sample [57,58].

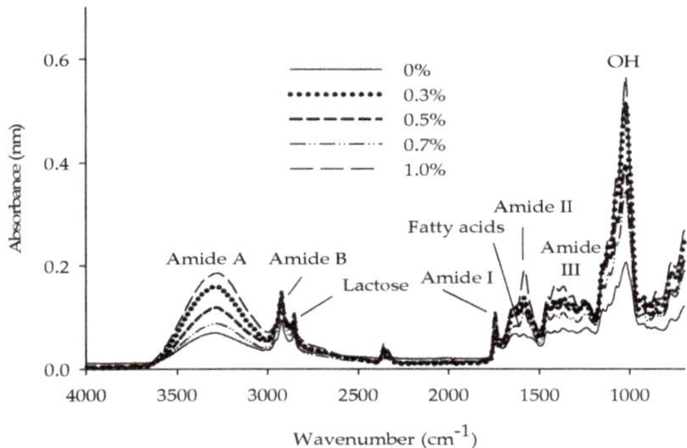

Figure 2. Fourier transform-infrared spectra of the functional whey beverage with different concentrations of hydrolyzed collagen.

3.5. Thermal Properties of Whey Beverage

Table 2 did not present significant difference ($p \geq 0.05$) in Tm between the different treatments and the control. However, significant differences ($p \leq 0.05$) were observed between the control treatment (0%) and the treatments with HC. The presence of HC in the beverage not only favors the thermal stability, but also helps to increase pH (Table 1). Fitzsimons et al. [59] showed the relation between the differential scanning calorimetry and the protein content increments, presence of salt, and pH in whey products. Heating treatment causes the denaturation of proteins and thus causes changes in the functional and structural characteristics of proteins [60]. Changes in pH could be affected by the thermal properties in whey products [61,62]. The presence of other components, such as HC, sucrose, and lactose-free milk powder, can also modify the composition of the beverage. Joyce et al. [63] reported similar results in milk formulas with different concentrations of whey protein: the product with a higher protein content presented a higher denaturation temperature.

Table 2. Enthalpy (ΔH) and melting temperature (T_m) of fermented whey beverages following thermal treatment at 60 °C for 30 min.

Hydrolyzed Collagen Concentration (%)	ΔH (J/g)	T_m (°C)
0	94.81 ± 6.01 [a]	163.504 ± 5.16 [a]
0.3	12.77 ± 4.23 [b]	158.88 ± 5.84 [a]
0.5	8.80 ± 6.07 [b]	147.03 ± 4.43 [a]
0.7	11.53 ± 2.26 [b]	165.612 ± 2.75 [a]
1.0	14.818 ± 3.37 [b]	161.4 ± 5.90 [a]

Values are means ± SD based on 3 observations. Means with different lowercase letters differ ($p \leq 0.05$).

3.6. Antioxidant Activity by ABTS and DPPH Radical Inhibition

Figure 3 shows that the highest radical inhibition for both ABTS and DPPH resulted in treatment with the highest HC (1%) with values of 48.30 ± 0.52% and 30.06 ± 1.09%, respectively. These values were different ($p \leq 0.05$) compared to the 0% (control) concentration. This antioxidant activity is related to the hydrophobic and aromatic amino acids, which can stabilize electron deficient radicals by donating protons [64]. However, the increase of antioxidant activity could be related to the addition of HC; when the HC concentration is increased, the radical inhibition is higher. The strong hydrolysis of collagen increased the concentration of isoleucine and methionine along the reduction of molecular weight and increased antioxidant activity [23]. Additionally, previous works have found that HC antioxidant activity is related to amino acid composition (tyrosine, histidine, and methionine) and low molecular weight. Isoleucine and methionine can be electron donors or hydrogens contributing to the increase of radical scavenging [65,66]. In beverages, the presence of bioactive peptides from both HC and milk whey not only contribute to their physicochemical stability, but also in the opportunity that these substances reach the consumer and promote their health [13]. Other works in the development of fermented whey beverages presented low ABTS radical inhibition compared to the inhibition obtained in this research [67,68]. In addition, DPPH radical inhibition was lower in beverages with orange, soursop, and carrot juice [69–71].

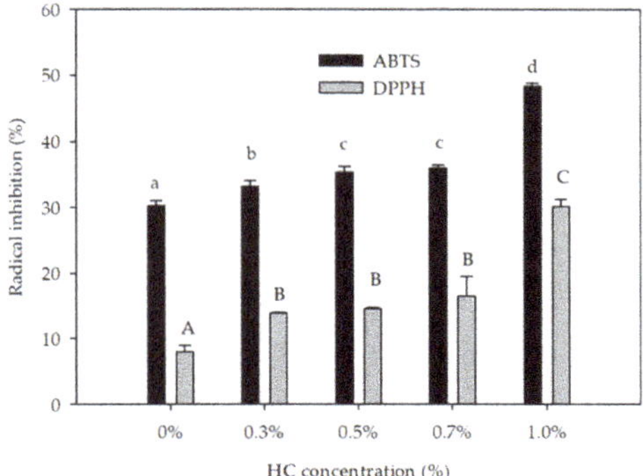

Figure 3. Antioxidant activity by ABTS (2,2′-Azino-bis(3-ethylbenzothiazoline-6-sulfonic acid)) and DPPH (2,2 diphenyl-1-picrylhydrazyl) radical inhibition of the whey beverage with different concentrations of added hydrolyzed collagen (HC). Values are means ± SD based on 3 observations. Means within the same category (i.e., ABTS or DPPH) with different letters differ ($p \leq 0.05$).

3.7. Microbiological Analysis

The beverage was pasteurized for 30 min at 60 °C. The main objective of this thermal treatment was the inhibition or elimination of microorganisms that can affect the quality of the food product. Thermal treatment was successful since no contamination by microorganisms were present (day 0). Additionally, the presence of the raw materials, such as whey and HC, could have a positive effect in inhibiting the presence of microorganisms. The antimicrobial property of the HC is related to the low molecular weight between 3–6 kDa and the presence of free amino acids [13,72]. In addition, milk whey is a good source of biological proteins, peptides, minerals, and vitamins, helping with the growth of LAB, and producing inhibitory metabolites that are antagonistic to pathogens [73]. The proteins present in milk whey, such as α-lacto albumin, β-lactoglobulin, or lactoferrin, are physiologically active and present antimicrobial and antiviral activities. However, the beverage with 0% HC after 15 and 30 days of storage at 4 °C was positive for yeast and mold. The treatments with HC did not show the presence of *Salmonella*, *E. coli*, molds, yeast, or anaerobic mesophiles. The HC, as a functional ingredient, could act as a natural antimicrobial to help extend the shelf-life of food. Previous works showed the presence of aerobic mesophiles, *E. coli*, mold, and yeast in an orange juice whey beverage [64], whey-based sports beverages [74], and a whey beverage with lutein [75]. However, when HC was added as a functional ingredient to different types of foods like soup [76], sausages [77], orange and apple juice with HC, dairy beverages [49], and fermented dairy beverages [48], lower or no presence of pathogenic microorganisms, such as *Salmonella* and aerobic mesophiles, were shown.

3.8. In Vitro Bioavailability

Bioavailability describes the amount of nutrients or functional ingredients that are actually absorbed, distributed to the tissues, metabolized, and eventually excreted by the body [78,79]. Figure 4 shows the in vitro bioavailability of HC present in the whey beverages. All the treatments resulted with significant differences ($p \leq 0.05$). Beverages with a 0.3% addition of HC showed the lowest rate of absorption at 1.22 ± 0.15%. When HC was increased, the rate of absorption showed the same behavior up to 58.95 ± 2.72% for the 1.0% treatment. HC was added as a functional ingredient into the whey beverage to modify the physicochemical properties to increase the functional properties, such as the antimicrobial and antioxidant properties already mentioned. The advantage of the HC used in this experiment is related to its low molecular weight (5.79 ± 0.30 KDa) and hydroxyproline concentration of about 17.85 ± 1.22 mg/L [20]. Peptides with low molecular weight, such as HC, are more bioavailable compared to native proteins [80–82]. When collagen is consumed in the form of a native protein, it needs to undergo degradation by gastric and pancreatic protease action, to become small peptides or free amino acids. Subsequently, peptides could be transported by the intestine in the form of free amino acids and di- or tripeptides, which results in poor assimilation. However, HC plays an important role because collagen in its hydrolyzed form can be easily assimilated by the body due to previous digestion with enzymes [83–86]. In addition, the presence of hydroxyproline promotes the major assimilation of HC because this amino acid along with proline can be absorbed in the gastrointestinal tract after ingestion [82]. Similar results were shown in a juice beverage with HC: the higher the HC concentration in the beverage, the higher the bioavailability was observed [30].

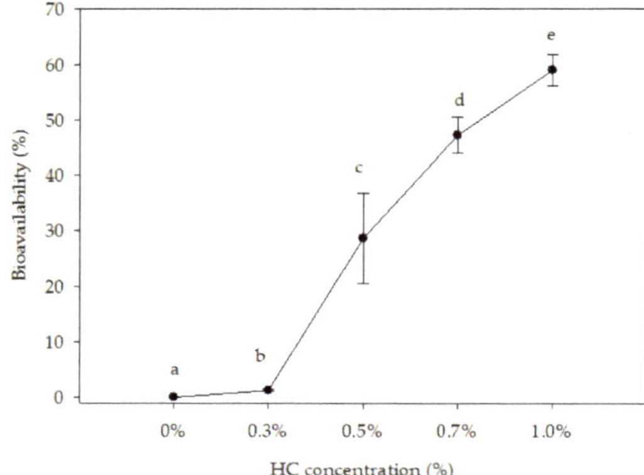

Figure 4. In vitro bioavailability of the hydrolyzed collagen (HC) present in the whey beverage. Values are means ± SD based on 3 observations. Means with different lowercase letters differ ($p \leq 0.05$).

4. Conclusions

From this study, it can be concluded that the presence of both HC and milk whey in a functional beverage helped to increase the nutritional value of the beverage due to the high protein concentration and no changes in the fat and lactose concentrations. HC played the role of an antioxidant ingredient and prevented the presence of some pathogenic microorganisms. Additionally, the in vitro bioavailability of HC demonstrated high rates of absorption due to its low molecular weight. These results could be of great importance for their application in the development of beverages oriented towards athletes and elderly people.

Author Contributions: Conceptualization: R.G.C.-M. and G.A.-Á.; Data curation: X.A.P.-M. and G.C.-L.; Formal analysis: A.L.-L. and Rafael G.C.-M.; Investigation: A.L.-L. and X.A.P.-M.; Methodology: A.L.-L. and G.C.-L.; Supervision: G.A.-Á.; Writing—Original draft: A.L.-L.; Writing—Review & editing: G.A.-Á. All authors have read and agreed to the published version of the manuscript.

Funding: CONACyT: grant number 621400.

Acknowledgments: To Dimitrios Zeugolis for his technical support during a research stay at the University of Galway, Ireland.

Conflicts of Interest: The authors declare no conflict of interest.

Abbreviations

ABTS	2,2'-Azino-bis (3-ethylbenzothiazoline-6-sulfonic acid)
ANOVA	analysis of variance
DPPH	2,2-diphenyl-1-picrylhydrazyl
FTIR	Fourier transform-infrared spectroscopy
HC	hydrolyzed collagen
LAB-	lactic acid bacteria

References

1. Bigliardi, B.; Galati, F. Innovation trends in the food industry: The case of functional foods. *Trends Food Sci. Technol.* **2013**, *31*, 118–129. [CrossRef]
2. Raman, M.; Ambalam, P.; Doble, M. 9-Probiotics, Prebiotics, and Fibers in Nutritive and Functional Beverages. In *Nutrients in Beverages*; Grumezescu, A.M., Holban, A.M., Eds.; Elsevier: Amsterdam, The Netherlands, 2019; pp. 315–367. [CrossRef]
3. Hasler, C.M. Functional Foods: Benefits, Concerns and Challenges—A Position Paper from the American Council on Science and Health. *J. Nutr.* **2002**, *132*, 3772–3781. [CrossRef]
4. Ofori, J.A.; Peggy, Y.-H. Novel Technologies for the Production of Functional Foods. In *Bio-Nanotechnology: A Revolution in Food, Biomedical and Health Sciences*; Shahidi, F., Bagchi, D., Bagchi, M., Moriyama, H., Shahidi, F., Eds.; Wiley-Blackwell: Hoboken, NJ, USA, 2013; pp. 141–162.
5. Özer, B.H.; Kirmaci, H.A. Functional milks and dairy beverages. *Int. J. Dairy Technol.* **2010**, *63*, 1–15. [CrossRef]
6. Turkmen, N.; Akal, C.; Ozer, B. Probiotic dairy-based beverages: A review. *J. Funct. Foods* **2019**, *53*, 62–75. [CrossRef]
7. Kareb, O.; Aïder, M. Whey and Its Derivatives for Probiotics, Prebiotics, Synbiotics, and Functional Foods: A Critical Review. *Probiotics Antimicrob. Proteins* **2019**, *11*, 348–369. [CrossRef] [PubMed]
8. Lee, H.; Song, M.; Hwang, S. Optimizing bioconversion of deproteinated cheese whey to mycelia of Ganoderma lucidum. *Process. Biochem.* **2003**, *38*, 1685–1693. [CrossRef]
9. Ásványi, B.; Reichart, O.; Szigeti, J.; Varga, L. Screening and selection of Kluyveromyces strains for use in batch production of single-cell protein from cheese whey. *Milchwissenschaft* **2006**, *61*, 378–381.
10. Alvarado Carrasco, C.; Guerra, M. Lactosuero como fuente de péptidos bioactivos. *Anales Venez. Nutr.* **2010**, *23*, 45–50.
11. Chatterjee, G.; De Neve, J.; Dutta, A.; Das, S. Formulation and statistical evaluation of a ready-to-drink whey based orange beverage and its storage stability. *Rev. Mex. Ing. Quím.* **2015**, *14*, 253–264.
12. Ramírez-Navas, J.S. Aprovechamiento Industrial de Lactosuero Mediante Procesos Fermentativos. *Publ. Investig.* **2012**, *6*, 69–83. [CrossRef]
13. Vivas, Y.A.; Morales, A.J.; Otálvaro, Á.M. Aprovechamiento de lactosuero para el desarrollo de una bebida refrescante con antioxidantes naturales. *Aliment. Hoy* **2017**, *24*, 185–199.
14. Pescuma, M.; Hebert, E.; Mozzi, F.; De Valdez, G.F. Whey fermentation by thermophilic lactic acid bacteria: Evolution of carbohydrates and protein content. *Food Microbiol.* **2008**, *25*, 442–451. [CrossRef] [PubMed]
15. Skov, K.; Oxfeldt, M.; Thøgersen, R.; Hansen, M.G.; Bertram, H.C. Enzymatic Hydrolysis of a Collagen Hydrolysate Enhances Postprandial Absorption Rate-A Randomized Controlled Trial. *Nutrients* **2019**, *11*, 1064. [CrossRef] [PubMed]
16. León-López, A.; Morales-Peñaloza, A.; Martínez-Juárez, V.M.; Vargas-Torres, A.; Zeugolis, D.I.; Aguirre-Álvarez, G. Hydrolyzed Collagen—Sources and Applications. *Molecules* **2019**, *24*, 4031. [CrossRef]
17. Aguirre-Cruz, G.; León-López, A.; Cruz-Gómez, V.; Alvarado, R.J.; Aguirre-Álvarez, G. Collagen Hydrolysates for Skin Protection: Oral Administration and Topical Formulation. *Antioxidants* **2020**, *9*, 181. [CrossRef]
18. Najafian, L.; Babji, A. A review of fish-derived antioxidant and antimicrobial peptides: Their production, assessment, and applications. *Peptides* **2012**, *33*, 178–185. [CrossRef]
19. Bradford, N. A rapid and sensitive method for the quantitation microgram quantities of a protein isolated from red cell membranes. *Anal. Biochem.* **1976**, *72*, 248–254. [CrossRef]
20. AOAC. *Official Methods of Analysis. pH Determination*, 18th ed.; AOAC: Gaithersburg, MD, USA, 1997.
21. AOAC. *Official Methods of Analysis. Fat Determination*, 17th ed.; AOAC: Gaithersburg, MD, USA, 2000.
22. Trevelyan, W.E.; Harrison, J.S. Studies on yeast metabolism. Fractionation and microdetermination of cell carbohydrates. *Biochem. J.* **1952**, *50*, 298–303. [CrossRef]
23. León-López, A.; Fuentes-Jiménez, L.; Fuentes, A.D.H.; Campos-Montiel, R.G.; Aguirre-Álvarez, G. Hydrolysed Collagen from Sheepskins as a Source of Functional Peptides with Antioxidant Activity. *Int. J. Mol. Sci.* **2019**, *20*, 3931. [CrossRef]
24. Tirado-Armesto, D.F. Elaboración de una bebida láctea a base de lactosuero fermentado usando *Streptococcus salivarius* ssp., Thermophilus y *Lactobacillus casei* ssp. casei. *Cienc. Tecnol. Aliment.* **2015**, *13*, 13–19.

25. López-Legarda, X.; Taramuel-Gallardo, A.; Arboleda-Echavarría, C.; Segura-Sánchez, F.; Restrepo-Betancur, L.F. Comparación de métodos que utilizan ácido sulfúrico para la determinación de azúcares totales. *Rev. Cuba. Quími.* **2017**, *29*, 180–198.
26. AOAC. *Official Methods of Analysis. Ashes Determination*, 16th ed.; AOAC: Washington, DC, USA, 1995.
27. AOAC. *Official Methods of Analysis. Amino Acids*, 17th ed.; AOAC: Gaithersburg, MD, USA, 2006.
28. Re, R.; Pellegrini, N.; Proteggente, A.; Pannala, A.; Yang, M.; Rice-Evans, C. Antioxidant activity applying an improved ABTS radical cation decolorization assay. *Free. Radic. Boil. Med.* **1999**, *26*, 1231–1237. [CrossRef]
29. Brand-Williams, W.; Cuvelier, M.; Berset, C. Use of a free radical method to evaluate antioxidant activity. *LWT Food Sci. Technol.* **1995**, *28*, 25–30. [CrossRef]
30. Bilek, S.E.; Bayram, S.K. Fruit juice drink production containing hydrolyzed collagen. *J. Funct. Foods* **2015**, *14*, 562–569. [CrossRef]
31. Poveda, E. Suero lácteo, generalidades y potencial uso como fuente de calcio de alta biodisponibilidad. *Rev. Chil. Nutr.* **2013**, *40*, 397–403. [CrossRef]
32. Skryplonek, K. The use of acid whey for the production of yogurt-type fermented beverages. *Mljek. Čas. Unapr. Proizv. Prerade Mlijek.* **2018**, *68*, 139–149.
33. Atallah, A.A.; Gemiel, D.G. Preparation of New Carbonated Beverages Based on Hydrolyzed Whey by Fruit and Some Herbs. *Am. J. Food Sci. Technol.* **2020**, *8*, 49–55.
34. Huertas, R.A.P. Lactosuero: Importancia en la industria de alimentos. *Rev. Fac. Nac. Agron. Medellín* **2009**, *62*, 4967–4982.
35. Jelen, P. Industrial whey processing technology: An overview. *J. Agric. Food Chem.* **1979**, *27*, 658–661. [CrossRef]
36. Smith, T.; Foegeding, E.; Drake, M. Flavor and Functional Characteristics of Whey Protein Isolates from Different Whey Sources. *J. Food Sci.* **2016**, *81*, C849–C857. [CrossRef]
37. Prazeres, A.R.; Carvalho, M.; Rivas, J.; Carvalho, F. Cheese whey management: A review. *J. Environ. Manag.* **2012**, *110*, 48–68. [CrossRef] [PubMed]
38. Macwan, S.R.; Dabhi, B.K.; Parmar, S.; Aparnathi, K. Whey and its Utilization. *Int. J. Curr. Microbiol. Appl. Sci.* **2016**, *5*, 134–155. [CrossRef]
39. Mann, B.; Kumari, A.; Kumar, R.; Sharma, R.; Prajapati, K.; Mahboob, S.; Athira, S. Antioxidant activity of whey protein hydrolysates in milk beverage system. *J. Food Sci. Technol.* **2014**, *52*, 3235–3241. [CrossRef] [PubMed]
40. Skryplonek, K.; Dmytrów, I.; Mituniewicz-Małek, A. Probiotic fermented beverages based on acid whey. *J. Dairy Sci.* **2019**, *102*, 7773–7780. [CrossRef] [PubMed]
41. Guimarães, P.M.; Teixeira, J.A.; Domingues, L. Fermentation of lactose to bio-ethanol by yeasts as part of integrated solutions for the valorisation of cheese whey. *Biotechnol. Adv.* **2010**, *28*, 375–384. [CrossRef] [PubMed]
42. Hailu, M.; Tola, A.; Teshome, G.; Agza, B. Development of Beverages from Traditional Whey and Natural Fruit Juices. *Food Sci. Nutr. Complet. Res.* **2019**, 85–93. Available online: http://publication.eiar.gov.et:8080/xmlui/bitstream/handle/123456789/3288/Food%20Science%20Proceedings-pdg.pdf?sequence=1&isAllowed=y#page=89 (accessed on 10 August 2020).
43. De Lima, A.V.S.C.; Nicolau, E.S.; Rezende, C.S.M.E.; Torres, M.C.L.; Novais, L.G.; Soares, N.R. Characterization and sensory preference of fermented dairy beverages prepared with different concentrations of whey and araticum pulp. *Semin. Ciênc. Agrár.* **2016**, *37*, 4011. [CrossRef]
44. Alves, A.T.S.E.; Spadoti, L.M.; Zacarchenco, P.B.; Trento, F.K.H.S.E.; Alves, A.S. Probiotic Functional Carbonated Whey Beverages: Development and Quality Evaluation. *Beverages* **2018**, *4*, 49. [CrossRef]
45. Saha, P.; Ray, P.; Ghatak, P.; Bag, S.; Hazra, T. Physico-chemical quality and storage stability of fermented Chhana whey beverages. *Indian J. Dairy Sci.* **2017**, *70*, 398–403.
46. Akpinar, A.; Torunoglu, F.A.; Yerlikaya, O.; Kinik, O.; Akbulut, N.; Uysal, H.R. Fermented probiotic beverages produced with reconstituted whey and cow milk. *Agro FOOD Ind. Hi Tech.* **2015**, *26*, 4.
47. Schrieber, R.; Gareis, H. *Gelatine Handbook: Theory and Industrial Practice*; John Wiley & Sons: Hoboken, NJ, USA, 2007.

48. Gerhardt, Â.; Monteiro, B.W.; Gennari, A.; Lehn, D.N.; de Souza, C.F.V. CARACTERÍSTICAS FÍSICO-QUÍMICAS E SENSORIAIS DE BEBIDAS LÁCTEAS FERMENTADAS UTILIZANDO SORO DE RICOTA E COLÁGENO HIDROLISADO Physicochemical and sensory characteristics of fermented dairy drink using ricotta cheese whey and hydrolyzed collagen. *Rev. Inst. Laticínios Cândido Tostes* **2013**, *68*, 41–50. [CrossRef]

49. Rigoto, J.D.M.; Ribeiro, T.H.; Stevanato, N.; Sampaio, A.R.; Ruiz, S.; Bolanho, B.C. Effect of açaí pulp, cheese whey, and hydrolysate collagen on the characteristics of dairy beverages containing probiotic bacteria. *J. Food Process. Eng.* **2019**, *42*, e12953. [CrossRef]

50. López, P.I.G.; Zambrano Ángela, M.Z.; Rosado, C.F.R.; Peña, A.M. Evaluación de una bebida láctea fermentada novel a base de lactosuero y harina de camote. *La Téc.* **2018**, *19*, 47–60.

51. Chairunnisa, H.; Balia, R.L.; Wulandari, E. Chemical and Microbiological Characteristics of Fermented Cheese Whey Beverages with Soymilk Powder Addition. In *IOP Conference Series: Earth and Environmental Science*; IOP Publishing: Bristol, UK, 2019; Volume 334, p. 012043.

52. Imbachí-Narváez, P.C.; Sepúlveda-Valencia, J.U.; Rodriguez-Sandoval, E. Effect of modified cassava starch on the rheological and quality properties of a dairy beverage prepared with sweet whey. *Food Sci. Technol.* **2019**, *39*, 134–142. [CrossRef]

53. Fanelli, S.; Zimmermann, A.; Tótoli, E.G.; Salgado, H.R.N. FTIR Spectrophotometry as a Green Tool for Quantitative Analysis of Drugs: Practical Application to Amoxicillin. *J. Chem.* **2018**, *2018*, 1–8. [CrossRef]

54. Andrade, J.; Pereira, C.G.; Junior, J.C.D.A.; Viana, C.C.R.; Neves, L.N.D.O.; Da Silva, P.H.F.; Bell, M.J.V.; Anjos, V.D.C.D. FTIR-ATR determination of protein content to evaluate whey protein concentrate adulteration. *LWT* **2019**, *99*, 166–172. [CrossRef]

55. Ferraro, V.; Madureira, A.R.; Sarmento, B.; Gomes, A.M.; Pintado, M.E. Study of the interactions between rosmarinic acid and bovine milk whey protein α-Lactalbumin, β-Lactoglobulin and Lactoferrin. *Food Res. Int.* **2015**, *77*, 450–459. [CrossRef]

56. Salari, M.; Khiabani, M.S.; Mokarram, R.R.; Ghanbarzadeh, B.; Kafil, H.S.; Khiabani, M.S. Preparation and characterization of cellulose nanocrystals from bacterial cellulose produced in sugar beet molasses and cheese whey media. *Int. J. Boil. Macromol.* **2019**, *122*, 280–288. [CrossRef]

57. Wang, T.; Tan, S.-Y.; Mutilangi, W.; Plans, M.; Rodriguez-Saona, L. Application of infrared portable sensor technology for predicting perceived astringency of acidic whey protein beverages. *J. Dairy Sci.* **2016**, *99*, 9461–9470. [CrossRef]

58. Arruda, H.S.; Silva, E.K.; Pereira, G.A.; Meireles, M.A.A.; Pastore, G.M. Inulin thermal stability in prebiotic carbohydrate-enriched araticum whey beverage. *LWT* **2020**, *128*, 109418. [CrossRef]

59. Fitzsimons, S.M.; Mulvihill, D.M.; Morris, E.R. Denaturation and aggregation processes in thermal gelation of whey proteins resolved by differential scanning calorimetry. *Food Hydrocoll.* **2007**, *21*, 638–644. [CrossRef]

60. Boostani, S.; Aminlari, M.; Moosavi-Nasab, M.; Niakosari, M.; Mesbahi, G. Fabrication and characterisation of soy protein isolate-grafted dextran biopolymer: A novel ingredient in spray-dried soy beverage formulation. *Int. J. Boil. Macromol.* **2017**, *102*, 297–307. [CrossRef] [PubMed]

61. Agyare, K.K.; Damodaran, S. pH-Stability and Thermal Properties of Microbial Transglutaminase-Treated Whey Protein Isolate. *J. Agric. Food Chem.* **2010**, *58*, 1946–1953. [CrossRef] [PubMed]

62. Li, X.; Gao, Z.; Li, T.; Sarker, S.-K.; Chowdhury, S.; Jiang, Z.; Mu, Z. Effects of pH Values on Physicochemical Properties and Antioxidant Potential of Whey Protein Isolate-safflower Yellow Complexes. *Food Sci. Technol. Res.* **2018**, *24*, 475–484. [CrossRef]

63. Joyce, A.M.; Kelly, A.L.; O'Mahony, J.A.; Brodkorb, A. Separation of the effects of denaturation and aggregation on whey-casein protein interactions during the manufacture of a model infant formula. *Dairy Sci. Technol.* **2016**, *96*, 787–806. [CrossRef]

64. Vieira, A.H.; Balthazar, C.F.; Guimaraes, J.T.; Rocha, R.S.; Pagani, M.M.; Esmerino, E.A.; Silva, M.C.; Raices, R.S.; Tonon, R.V.; Cabral, L.M.; et al. Advantages of microfiltration processing of goat whey orange juice beverage. *Food Res. Int.* **2020**, *132*, 109060. [CrossRef]

65. Chen, H.-M.; Muramoto, K.; Yamauchi, F.; Fujimoto, K.; Nokihara, K. Antioxidative Properties of Histidine-Containing Peptides Designed from Peptide Fragments Found in the Digests of a Soybean Protein. *J. Agric. Food Chem.* **1998**, *46*, 49–53. [CrossRef]

66. Pownall, T.L.; Udenigwe, C.C.; Aluko, R.E. Amino Acid Composition and Antioxidant Properties of Pea Seed (*Pisum sativum* L.) Enzymatic Protein Hydrolysate Fractions. *J. Agric. Food Chem.* **2010**, *58*, 4712–4718. [CrossRef]

67. Arranz, E.; Corrochano, A.; Shanahan, C.; Villalva, M.; Jaime, L.; Santoyo, S.; Callanan, M.; Murphy, E.; Giblin, L. Antioxidant activity and characterization of whey protein-based beverages: Effect of shelf life and gastrointestinal transit on bioactivity. *Innov. Food Sci. Emerg. Technol.* **2019**, *57*, 102209. [CrossRef]

68. Arsić, S.; Bulatović, M.; Zarić, D.; Kokeza, G.; Subić, J.; Rakin, M. Functional fermented whey carrot beverage-qualitative, nutritive and techno-economic analysis. *Rom. Biotechnol. Lett.* **2018**, *23*, 13496–13504.

69. Sady, M.; Jaworska, G.; Grega, T.; Bernas, E.; Domagala, J. Application of acid whey in orange drink production. *Food Technol. Biotechnol.* **2013**, *51*, 266.

70. Guimarães, J.T.; Silva, E.K.; Ranadheera, C.S.; Moraes, J.; Raices, R.S.; Silva, M.C.; Ferreira, M.S.; Freitas, M.Q.; Meireles, M.A.A.; Cruz, A.G. Effect of high-intensity ultrasound on the nutritional profile and volatile compounds of a prebiotic soursop whey beverage. *Ultrason. Sonochem.* **2019**, *55*, 157–164. [CrossRef] [PubMed]

71. Ferreira, M.V.S.; Cappato, L.P.; Silva, R.; Rocha, R.S.; Guimarães, J.T.; Balthazar, C.F.; Esmerino, E.A.; Freitas, M.Q.; Rodrigues, F.N.; Granato, D.; et al. Ohmic heating for processing of whey-raspberry flavored beverage. *Food Chem.* **2019**, *297*, 125018. [CrossRef]

72. Gómez-Guillén, M.C.; Giménez, B.; López-Caballero, M.; Montero, M. Functional and bioactive properties of collagen and gelatin from alternative sources: A review. *Food Hydrocoll.* **2011**, *25*, 1813–1827. [CrossRef]

73. Singh, D.; Vij, S.; Singh, B.P. Antioxidative and antimicrobial activity of whey based fermented soy beverage with curcumin supplementation. *Indian J. Dairy Sci.* **2016**, *69*, 171–177.

74. Valadao, N.K.; De Andrade, I.M.G. Development of a Ricotta Cheese Whey-based Sports Drink. *Adv. Dairy Res.* **2016**, *4*, 156. [CrossRef]

75. Rocha, J.D.C.G.; Mendonça, A.C.; Viana, K.W.C.; Maia, M.D.P.; De Carvalho, A.; Minim, V.P.R.; Stringheta, P.C. Beverages formulated with whey protein and added lutein. *Ciênc. Rural* **2017**, *47*. [CrossRef]

76. Benjakul, S.; Chantakun, K.; Karnjanapratum, S. Impact of retort process on characteristics and bioactivities of herbal soup based on hydrolyzed collagen from seabass skin. *J. Food Sci. Technol.* **2018**, *55*, 3779–3791. [CrossRef]

77. Sousa, S.C.; Fragoso, S.P.; Penna, C.R.; Arcanjo, N.M.; Silva, F.A.; Ferreira, V.C.; Barreto, M.D.; Araújo, B.S. Quality parameters of frankfurter-type sausages with partial replacement of fat by hydrolyzed collagen. *LWT Food Sci. Technol.* **2017**, *76*, 320–325. [CrossRef]

78. Holst, B.; Williamson, G. Nutrients and phytochemicals: From bioavailability to bioefficacy beyond antioxidants. *Curr. Opin. Biotechnol.* **2008**, *19*, 73–82. [CrossRef]

79. Lutz, M. Biodisponibilidad de compuestos bioactivos en alimentos. *Perspect. Nutr. Hum.* **2013**, *15*, 217–226.

80. Hajirostamloo, B. Bioactive component in milk and dairy product. *World Acad. Sci. Eng. Technol.* **2010**, *72*, 162–166.

81. Feng, M.; Betti, M. Transepithelial transport efficiency of bovine collagen hydrolysates in a human Caco-2 cell line model. *Food Chem.* **2017**, *224*, 242–250. [CrossRef] [PubMed]

82. Sontakke, S.B.; Jung, J.-H.; Piao, Z.; Chung, H.J. Orally Available Collagen Tripeptide: Enzymatic Stability, Intestinal Permeability, and Absorption of Gly-Pro-Hyp and Pro-Hyp. *J. Agric. Food Chem.* **2016**, *64*, 7127–7133. [CrossRef] [PubMed]

83. Clemente, A. Enzymatic protein hydrolysates in human nutrition. *Trends Food Sci. Technol.* **2000**, *11*, 254–262. [CrossRef]

84. Egerton, S.; Culloty, S.; Whooley, J.; Stanton, C.; Ross, R.P. Characterization of protein hydrolysates from blue whiting (Micromesistius poutassou) and their application in beverage fortification. *Food Chem.* **2018**, *245*, 698–706. [CrossRef]

85. Ganapathy, V.; Gupta, N.; Martindale, R.G. Protein Digestion and Absorption. In *Physiolog of the Gastrointestinal Tract*; Elsevier: Amsterdam, The Netherlands, 2006; pp. 1667–1692.

86. Moskowitz, R.W. Role of collagen hydrolysate in bone and joint disease. *Semin. Arthritis Rheum.* **2000**, *30*, 87–99. [CrossRef]

© 2020 by the authors. Licensee MDPI, Basel, Switzerland. This article is an open access article distributed under the terms and conditions of the Creative Commons Attribution (CC BY) license (http://creativecommons.org/licenses/by/4.0/).

 *foods*

Article

Ripening of Nostrano Valtrompia PDO Cheese in Different Storage Conditions: Influence on Chemical, Physical and Sensory Properties

Luca Bettera [1], Marcello Alinovi [1,*], Roberto Mondinelli [2] and Germano Mucchetti [1]

[1] Food and Drug, University of Parma, Parco Area delle Scienze, 47/A 43124 Parma, Italy; luca.bettera@unipr.it (L.B.); germano.mucchetti@unipr.it (G.M.)
[2] Consorzio di Tutela del formaggio Nostrano Valtrompia, Via G. Matteotti, 327, 25063 Gardone Val Trompia, Italy; mondinellirob@gmail.com
*   Correspondence: marcello.alinovi@studenti.unipr.it; Tel.: +39-334-328-3206

Received: 21 July 2020; Accepted: 8 August 2020; Published: 12 August 2020

Abstract: Nostrano Valtrompia is a hard, long-ripened, Italian Protected Designation of Origin (PDO) cheese typically produced by applying traditional cheesemaking practices in small dairies. Due to the limited production, this cheese is characterized by an important market price. Nostrano Valtrompia physico-chemical and sensory quality can be influenced by the duration and conditions of ripening. The objectives of this work were to characterize the physico-chemical and sensory characteristics of Nostrano Valtrompia cheese ripened for 12 and 16 months and to study the influence of different ripening warehouses: a temperature conditioned warehouse (TCW) and in a traditional, not conditioned warehouse (TNCW). The moisture gradient from the rind to the center of the cheese influenced texture, moisture, aw and color. Ripening in different warehouses did not affect the overall appreciation of the cheese nor other physico-chemical (color, moisture) or sensory traits. TCW cheeses were characterized by a slightly softer texture, slightly different openings distribution, and a different sensory perception than TNCW cheeses. These minor differences were related to the less variable environmental ripening conditions of TCW than TNCW. The results of this study can be useful to support the management of the ripening conditions of Nostrano Valtrompia PDO cheese and to rationally introduce new, suitable ripening sites.

Keywords: hard cheese; long ripened cheese; ripening rooms; environmental ripening conditions; quantitative descriptive analysis; texture; water activity; image analysis; cheesemaking technology

---

1. Introduction

Nostrano Valtrompia is a registered PDO cheese [1] included into the group of the Italian hard long ripened cheeses produced in the region of the Alps in Northern Italy, including, e.g., Fontina from Valle d'Aosta; Ossolano from Piemonte; Silter, Bagoss and Formai de Mut from Lombardia; Monte Veronese, Piave and Asiago from Veneto; Stelvio from Trentino Alto Adige and Montasio from Friuli Venezia Giulia [2,3].

While the production of the largest part of some of the cited cheeses is today made in industrial sites with modern equipment (e.g., Asiago, Montasio, Fontina, Stelvio or Piave), Nostrano Valtrompia cheese is still manufactured in small dairies that transform a maximum of 300 kg of milk per day, by applying traditional cheesemaking practices and by the use of traditional equipment [4,5].

The cheese is characterized by a low moisture content (max 36%), a fat-to-protein ratio usually lower than 0.85, a hard but not crumbly texture and the presence of small irregular openings in the structure mainly due to microbial fermentation [4,5].

The key points of Nostrano Valtrompia cheesemaking technology may be summarized as follows: (i) use of partially skimmed raw milk obtained from up to 4 milking (approximately 300 kg of milk) to produce wheels complying with the limitations in size provided by the standard (8–18 kg); (ii) production of one cheese per vat (rarely two), leading to a strong relation between milk amount and size of the cheese; (iii) use of copper vats with single wall heated by fire (from wood or gas burning); (iv) addition of saffron to milk or cheese curd grains (from 0.05 g to 0.2 g/100 kg milk) before cooking; (v) cooking of cheese curd granules up to 47–52 °C; (vi) minimum ripening time of 12 months.

The use of an autochthonous natural starter is frequent but not mandatory.

These fundamentals, common to Bagoss cheese produced in a near geographical area [6], may be considered for many aspects as the living memory of the ancient way of making Parmigiano Reggiano cheese before 1900 [7]. An example of how a living memory can become an innovation tool is the case of the use of saffron, an old practice left by Parmigiano Reggiano cheesemakers [8] but kept alive by some Italian cheeses as Nostrano Valtrompia, Bagoss or Piacentinu Ennese [6,9]. The addition of saffron to cheese is today sometimes presented as an innovative practice for ovine dairy products designed for niche markets [10] or to introduce healthy claims for consumers [11].

As a consequence, a better understanding of Nostrano Valtrompia cheese properties may be helpful to evaluate the choices made more than one hundred years ago by Parmigiano Reggiano dairies leading to the actual worldwide success but also to test today alternative ways to preserve this cultural heritage and to maintain the income of these small dairies, producing a cheese with an important market price (more than 22 Euro/kg at the dairy), higher than that of Parmigiano Reggiano cheese, able to repay the strong efforts required to make it.

Cheese characterization is of great interest in order to fulfill the definition of identity, authenticity and quality on the application of policies of the European Union and of the protection and information of consumers [12]. The recent recognition of the PDO status had the effect of creating a growing demand for Nostrano Valtrompia PDO cheese. However, few studies have described the physicochemical and sensory properties of this cheese [4,5].

The ripening method is another distinctive characteristic of Nostrano Valtrompia. Farmers-cheesemakers are rarely also "affineurs", keeping the ancient separation between cheesemakers and "affineurs" typical of the Parmigiano Reggiano area where an important part of the cheese is up to date ripened in large warehouses out of the dairies, storing more than 100,000 wheels. At the same time, the ripening represents a critical step in the Nostrano Valtrompia production. The limited storage capacity within the PDO area cannot satisfy the quantity of cheese produced, considering the minimum ripening period of 12 months imposed by the PDO regulation. This is a limiting factor for the growth of the Nostrano Valtrompia system.

Nostrano Valtrompia cheese ripening is characterized by the traditional practice of coating the surface with linseed oil. Coating is aimed to control both the rate of moisture evaporation and to limit the mold growth [13]. About every ten days, the cheese surface is cleaned with a knife from the film of linseed oil, polymerized because of air contact [14], and then newly oiled.

The large variability of cheese size [1] is responsible for a different evolution of the weight loss caused by moisture evaporation, affecting the a_w value gradient and the rate of biochemical reactions.

At present, the knowledge about the moisture distribution of hard long-time ripened cheeses is poor, as is that of textural and color properties, partly governed by the moisture content.

The aim of the paper is twofold: (i) analyze some physicochemical and sensory properties of Nostrano Valtrompia PDO cheese useful for its characterization; (ii) investigate whether these properties are affected by different ripening conditions. To reach this goal, cheeses were ripened for 12 and 16 months in a not conditioned traditional cellar and in a temperature conditioned warehouse and were characterized from a physico-chemical, sensory point of view.

2. Materials and Methods

2.1. Experimental Design

Experimental design is schematized in Figure 1. Four dairies (1 to 4) producing from a maximum of 2 cheeses/day to a minimum of 1 cheese every 3 days were chosen to represent the system of hard cheese production in Trompia Valley. Cheeses considered in this study were produced in the period from May to June 2017.

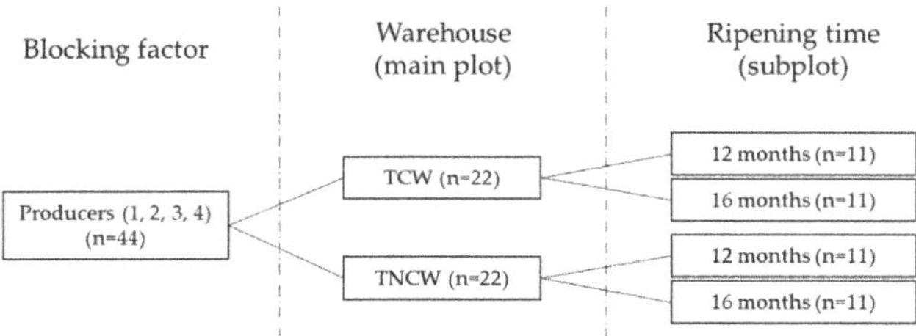

Figure 1. Experimental design schematization. TCW = temperature conditioned warehouse; TNCW = traditional non conditioned warehouse; *n* = number of cheeses.

The cheeses were randomly assigned and ripened for 12 and 16 months in two different warehouses: (i) temperature conditioned warehouse (TCW) (Formaggi Tre Valli, San Colombano, BS, Italy); (ii) traditional, not conditioned warehouse (TNCW) (Azienda Agricola Mauro Beltrami, Marmentino, BS, Italy).

2.2. Cheese Making

Dairies followed the cheese-making technology described by PDO standard of production. All the dairies used raw milk partially skimmed by spontaneous creaming and added saffron (~0.1 g/100 kg milk). The amount of milk used has been 184 ± 57 kg derived from one to three milkings.

The main features of the process are reported in Table 1. Values of pH and temperature during processing operations were measured with a Portamess pH-meter mod. 913 (Knick Elektronische, Berlin, Germany), equipped with a Double Pore F electrode (Hamilton Company, Reno, NV, USA) and a pt-1000 temperature probe.

Table 1. Duration of key Nostrano Valtrompia cheese manufacture steps and relative values of pH, temperature (expressed in °C) of milk and curd at the end of each cheese making step.

Cheese Making Step	Duration (min)		T(°C)		pH	
	Mean	±SD	Mean	±SD	Mean	±SD
Milk in cheese vat	N.A.	N.A.	20.75	2.39	6.51	0.10
Heating until coagulation	32.40	11.49	38.07	0.90	6.39	0.08
Coagulation (from rennet addition to gel cutting)	44.90	21.09	37.56	2.07	6.32	0.11
Cooking	19.00	7.80	50.06	1.70	6.11	0.19
Curd resting under whey	40.33	17.44	49.89	1.49	5.93	0.33

Raw milk was initially heated into copper vats using direct wooden fire until reaching a temperature of 38.1 ± 0.9. Two out of the four dairies (producers 2 and 4) used a low dosage (~0.2–0.5% v/v) of natural whey starters, while the other two (producers 1 and 3) did not use any starter. Milk was coagulated by adding powdered calf rennet (1:125,000 Soxhlet units).

After coagulation, the coagulum was cut using a blade and a whisk by producers 1 and 2, or only with the whisk by producers 3 and 4. After cutting, the cheese curd grains were cooked for a variable amount of time (19 ± 8 min) until reaching a final temperature of 50 ± 2 °C.

After cooking, the curd rested for 40 ± 17 min under hot whey, in order to allow for the reciprocal junction of the cheese curd particles and to favor whey syneresis, and it was finally extracted and molded with a linen cloth. Molded curds were overturned for two times to complete whey drainage and they were then dry salted a variable period from 2 to 11 days, depending on cheese weight and the choice of each cheesemaker.

Cheeses were stored after salting for a 7-days storage period in TCW warehouse and then half of them were addressed to the other ripening warehouse TNCW as reported in Section 2.1.

2.3. Temperature and Relative Humidity Monitoring during Ripening

During cheeses ripening, temperature and relative humidity (RH%) of the warehouses were continuously monitored with data loggers (Hygrolock, Rotronic, Hauppauge, NY, USA). Data were acquired with software HW4 (Rotronic, USA) during the ripening period with 1 h timesteps.

2.4. Cheese Physico-Chemical Characteristics

2.4.1. Cheese Sampling

To perform analyses after 12 and 16 months of ripening, cheese wheels were classified and portioned in representative parts of the wheels (slices of ~1.5 kg) (Figure 2a), they were under-vacuum packaged and stored at 4 ± 1 °C before being analyzed. The cheese slices were divided into three zones (Figure 2d): rind, defined as the external part of cheese with a depth of 0.5 cm; underrind, defined as the part of cheese with a depth between 0.5 cm and 1.5 cm and inner part of the cheese. This subdivision was made in order to highlight the possible influence of the two different ripening conditions (TCW, TNCW) and times (12, 16 months) in the different locations of the cheese. While image analysis features and sensory properties were measured only in the inner part of the cheese, the other physico-chemical parameters (water activity, moisture content, textural features, color) were assessed in the three different zones. Furthermore, textural and color measurements were performed in specific points of the slice within each of the three zones (Figure 2b,c). The distance of the points was normalized to 100 in function of the slice length, height (for underrind and inner part) and thickness (for the rind).

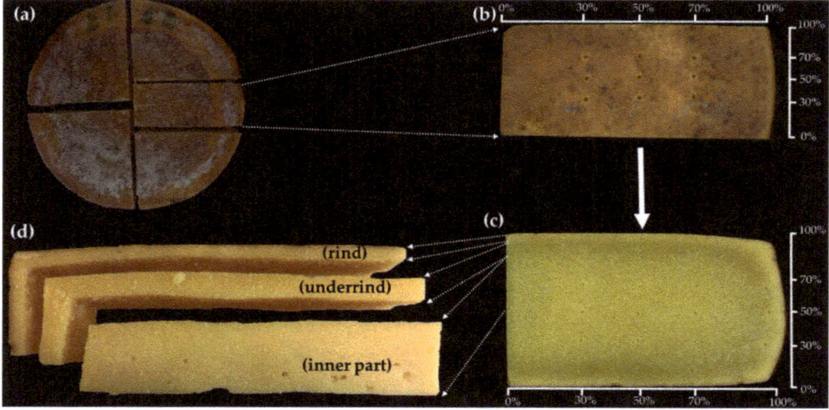

Figure 2. Cheese sampling and analysis schematization. (**a**) cheese wheel seen from the top: portioning of a representative slice; (**b**) focus on the rind: specific textural and color analysis points at 30–50–70% of slice length and thickness; (**c**) focus on the underrind and inner part: specific textural and color analysis points at 30–50–70% of slice length and height; (**d**) cheese zones subdivision: rind, underrind (part with a depth between 0.5 cm and 1.5 cm) and inner part.

2.4.2. Moisture Content and Water Activity

Moisture content of cheeses was measured by oven-drying samples at 102 °C [15] until a constant weight was reached. The measurements were performed in triplicate.

Water activity (a_w) of cheese samples was measured at 25 °C using an AquaLab Water Activity Meter Series 3TE with internal temperature control (Decagon Devices, Inc., Pullman, WA, USA).

2.4.3. Colorimetric Characteristics

Color of the different zones of cheese was measured using a Minolta Colorimeter (CM 2600d, Minolta Co., Osaka, Japan) equipped with a standard illuminant D65. The assessments were carried out at room temperature (25 °C). CIELAB color space was considered, and the parameters L^* (lightness, black = 0, white = 100), a^* (redness > 0, greenness < 0), b^* (yellowness > 0, blue < 0) were determined for each sample.

2.4.4. Textural Properties

Texture analysis of the cheeses was performed by means of a TA.XTplus Texture Analyzer (Stable Micro Systems, Godalming, UK) equipped with a 30 kg load cell and a stainless steel cylindrical probe with a 3 mm diameter (SMS P/3, Stable Micro Systems) according to a previously described penetration test [16]. Young's modulus, stress and strain at fracture were calculated according to Hort and Le Grys [17], from true strain (ε) and true stress (δ) parameters [18] (Equations (1) and (2)):

$$\varepsilon = ln(h_0/(h0 - \Delta h)) \tag{1}$$

$$\delta(t) = F(t)/A(t) \tag{2}$$

where h_0 is the original height (m), Δh represent the change in height of the sample (m), $F(t)$ is the force at time (t), and $A(t)$ is the surface area at time (t).

2.4.5. Image analysis

Image analysis of cheese slices was performed to estimate the presence of openings ("eyes" and/or "cracks") of the paste. Images of slices of cheese samples (0.21 × 0.30 mA4 scanner size) were acquired using a Hewlett Packard Scanjet 8200 scanner (Palo Alto, CA, USA) with a resolution of 600 dpi (corresponding to 236 pixels cm^{-1}) and saved in TIFF format. A black background was used to enhance the contrast of acquired images. Images were preprocessed by adjusting the parameters of brightness, contrast and gamma using a software (XnConvert 1.73, XnSoft Corp., Reims, France). Image analysis was performed with Matlab® R2018 b software (The MathWorks Inc., Natick, MA, USA), using the three following applications: (i) Color Thresholder to select the proper ranges of RGB (red, green, blue) channels of the color space that were applied to every cheese image to create a binary representation of the features-of-interest, according to Caccamo et al. [19]. Chosen values were: R: min = 0–max = 255,000; G: min = 0–max = 76,000; B: min = 0–max = 255,000; (ii) Image Segmenter to crop the area of cheese sample and modify the image in binary scale; (iii) Image Region Analyzer, to estimate the porosity and the shape and size of the openings. Porosity of each opening was calculated according to the following Equation (3):

$$Porosity\ (\%) = (Area\ of\ openings/Total\ area\ of\ image) \times 100 \tag{3}$$

According to preliminary trials, "eyes" were differentiated from "cracks" on the basis of the parameter eccentricity (e) of the ellipse that has the same second-moments as the region, as calculated by the software according to the following Equation (4):

$$e = c/a \tag{4}$$

where c is the distance between the foci of the ellipse, and a is its major axis length. The value is between 0 (the ellipse is a circle) and 1 (the ellipse is a line segment); openings having an eccentricity value lower of 0.9 were classified as "eyes"; openings with a value equal or higher than 0.9 were classified as "cracks".

To evaluate the size distribution of openings, the 10th, 50th and 90th percentiles (D10, D50 and D90) were calculated. To summarize the distribution, the span value and the mean size of openings were also calculated. The span value was calculated according to Equation (5):

$$Span = (D90 - D10)/D50 \tag{5}$$

2.5. Sensory Properties

Sensory properties of cheeses at 12 and 16 months of ripening were evaluated by means of a discrimination triangle test and by a consumer test. Tests were performed by comparing each time the two cheeses of the same dairy, stored in the two different warehouses.

2.5.1. Discriminant Test

To perform the discriminant triangle tests, workers and students of the University of Parma were chosen as panelists. The panel group was untrained but had previous experience with sensory analysis. Three samples (cubes of approximately 10 g) were placed in white plastic dishes and the panel group (at least 12 members for each sample, with a minimum and maximum age of 18 and 44, respectively) was asked to identify the odd cheese sample. A randomized three-digit code was provided to the panelists and used to identify and mark the samples. Moreover, the panel was also asked to indicate one or more particular sensory attributes responsible for any perceived difference between the samples. A total of 164 tests was performed.

2.5.2. Descriptive Test

The consumer test was performed in the geographical production area of Nostrano Valtrompia PDO cheese, in order to involve usual consumers of this cheese. Consumers were asked to taste two cheeses of the same producer stored in the different warehouses and to give a score for each sensory parameter considered using a 1–7 ordinal scale, where 1 represents the lack of the attribute, 7 the excessive presence of the attribute, and 4 is the optimal value. Consumers (162 women and 243 men) evaluated the cheeses in terms of firmness, presence of openings and eyes, color, aroma and taste. Finally, consumers were also asked to give an overall evaluation of cheese appreciation, using a 1–7 ordinal scale, where 1 and 7 represent the minimum and the maximum appreciation, respectively. The samples were cut into cubes of approximately 10 g and were placed in white plastic dishes; in order to judge the presence of openings and eyes, a slice of the cheese sample was showed to the consumers.

In total, 405 independent sensory evaluations were performed by consumers for cheeses ripened in TCW and 405 evaluations for cheeses ripened in TNCW.

2.6. Statistical Analysis

In the case of moisture, a_w, textural and colorimetric parameters, the effects of ripening warehouse and cheese zone were evaluated according to Alinovi et al. [20,21] by developing split plot ANOVA models. The effect of warehouse (W_i, $i = 1, 2$) was analyzed in whole-plot, and the different cheese producer (P_j, $j = 1, 2, 3, 4$) was used as the blocking factor; in the subplot the effects of ripening time (Rt_k, $k = 1, 2$) and of warehouse × cheese zone ($W \times Rt$) were analyzed (Equation (6)):

$$Y_{ijkl} = \mu + W_i + P_j + \delta_{ij} + Rt_k + (W \times Rt)_{ik} + \gamma_{ijk} \tag{6}$$

where δ_{ij}, γ_{ijk}, were the main plot and the subplot error terms, respectively, and Y_{ijkl} was the selected response variable.

The effect of the different warehouses on image analysis and consumer test sensory attributes was tested by performing one-way ANOVA models.

Split plot and one-way ANOVA models were built using PRC GLM of SAS (SAS Inst. Inc., Cary, NC, USA); lsmeans with LSD adjustment was used to perform multiple comparisons among means.

For discrimination triangle test, a binomial test was carried out in order to assess if the number of panel's correct classification responses give a higher probability level (p) than a random classification process ($p > 1/3$, $\alpha = 0.05$).

3. Results and Discussion

3.1. Warehouses Environmental Conditions

Among the several factors influencing the cheese ripening, temperature and relative humidity (RH%) of the warehouses play an important role in regulating the mass transfer between the cheese and the environment and the biochemical reactions rate.

By monitoring the environmental conditions, it was possible to reveal important differences between the two warehouses, as reported in Figure 3. TNCW was characterized by higher temperatures (13.4 ± 3.3 °C, compared to 10.1 ± 0.7 °C of TCW) and in general by a large temperature variation during the whole ripening period (Figure 3a); this was caused by the seasonal temperature variations and the absence of temperature control in the case of TNCW. In TNCW, temperatures reached higher peaks of 21.3 °C during the summer and lower peaks of 5.9 °C during the winter.

Moreover, TNCW showed also larger variations of temperature in a short time scale, that were due to the daily thermal excursion between day and night (Figure 3b).

On the other hand, temperature of TCW was quite constant over the year, and it did not suffer of daily temperature variations, thanks to the room temperature control (Figure 3a,b).

Concerning RH%, the two ripening rooms showed similar mean values and a similar high variability during 1-year period (89.6 ± 4.8% for TCW and 91.3 ± 5.0% for TNCW), that can also be observed in Figure 3c. It is important to note that strong RH% variations were present also in the case of TCW, since for this room, temperature was controlled, but there was no control of relative humidity. The RH% in the two warehouses was largely overlapped during the 1-year period, despite a clear deviation being observable from May 2018, with TNCW showing higher RH% (~95%) than TCW (~85%); this divergence continued until the end of the ripening period (result not shown).

In the case of TNCW, the lack of temperature control led to a small humidity variation that could be observed in a short time scale (Figure 3d) and that followed the daily thermal excursion observed in Figure 3b. This short-scale variation of RH% was not present in the case of TCW.

In general, the mean RH% value of the warehouses, which was found to be around 90 ± 5%, was slightly higher than the values reported for other Italian PDO hard or semi-hard, long ripened cheeses. Ripening conditions reported for Parmigiano-Reggiano cheese showed RH values from 74% to 85% [22] or more recently around 85% [3], while Asiago d'Allevo cheese is ripened at RH% of 80 ± 5% [23]. Conversely, the measured mean temperature of ripening of Nostrano Valtrompia, that is around 12 ± 3 °C, was found to be slightly lower to the one reported for the Parmigiano Reggiano cheese and in the same range of those cited for Asiago d'Allevo and Montasio [3,24–27]. Temperature and relative humidity reported in our study were in line with those previously reported by Mucchetti et al. [4], which monitored the environmental conditions for Nostrano Valtrompia ripened in both temperature-controlled and non-controlled warehouses.

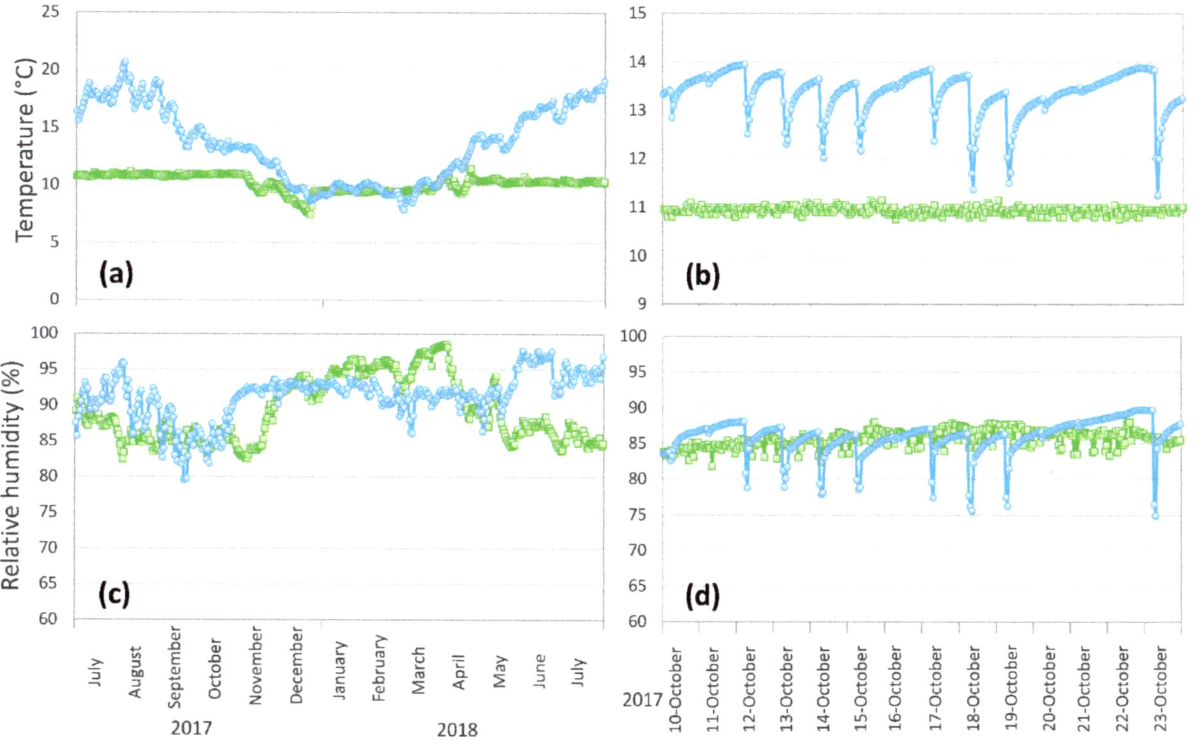

Figure 3. Temperature and relative humidity of Nostrano Valtrompia warehouses: temperature not conditioned warehouse (TNCW) (○) and temperature conditioned cellar (TCW) (□). Panels: (**a**) temperature variations over 1-year ripening period (July 2017–July 2018); (**b**) temperature variations over 2-week ripening period (10–23 October 2017); (**c**) relative humidity variations over 1-year ripening period (July 2017–July 2018); (**d**) relative humidity variations over 2-week ripening period (10–23 October 2017).

3.2. Moisture and Water Activity of the Cheese

Moisture content and water activity results are reported in Table 2.

Table 2. Moisture content and water activity (a_w) of Nostrano Valtrompia cheeses ripened for 12 and 16 months in temperature conditioned warehouse (TCW) and traditional non conditioned warehouse (TNCW).

Warehouse	Ripening Time (Months)	Moisture % (w/w) Mean ± SD			Water Activity (a_w) Mean ± SD		
		Rind	Underrind	Inner part	Rind	Underrind	INNER PART
TCW	12	16.1 [a] ± 2.0	24.7 [a] ± 2.1	32.0 [a] ± 1.2	0.882 [a] ± 0.012	0.919 [a] ± 0.009	0.931 [a] ± 0.006
	16	14.4 [a] ± 1.0	22.7 [b] ± 1.4	30.7 [b] ± 0.9	0.855 [b] ± 0.017	0.904 [b] ± 0.011	0.922 [b] ± 0.009
TNCW	12	15.1 [a] ± 1.3	24.6 [a] ± 1.3	30.7 [b] ± 1.1	0.872 [ab] ± 0.008	0.908 [c] ± 0.007	0.922 [b] ± 0.006
	16	15.6 [a] ± 1.2	24.5 [a] ± 1.5	30.5 [b] ± 1.0	0.860 [b] ± 0.014	0.895 [c] ± 0.010	0.909 [c] ± 0.007

[a–c] lowercase letters indicate significant differences ($p < 0.05$) within the same column.

In ripened cheeses, moisture gradients are present in the wheel [16]. As expected, the different zones of the cheese showed clear differences for both measured parameters; these differences reflected the moisture loss and migration phenomena involved during the ripening time of hard, long-ripened cheeses as Nostrano Valtrompia. The moisture content of the different cheese zones was found to be slightly lower than moisture data of Parmigiano Reggiano PDO cheese, which was found to have a moisture of 32.0 ± 1.4%, 26.7 ± 2.3% and 18.0 ± 2.2%, for inner part, underrind and rind at 12 months of ripening time, respectively [16]. These values were found to be in the range of those reported for Gouda cheese, despite the inner–outer moisture difference in Nostrano Valtrompia being higher [28]. In particular, the lower moisture content of Nostrano Valtrompia in the outer part of the cheese may indicate a stronger moisture loss; as the environmental conditions were similar or even less favorable for moisture loss than in the case of Parmigiano-Reggiano, the slightly lower moisture content of Nostrano Valtrompia can be a consequence of the higher surface to volume ratio and smaller cheese wheel dimensions if compared to Parmigiano-Reggiano [2,3,25].

As it is possible to observe, TCW cheeses showed a significant moisture loss between 12 and 16 months of ripening time in the case of the underrind and of the inner part of the cheese (−2.0% and −1.3%, respectively). On the contrary, TNCW did not show a significant decrease of moisture between 12 and 16 months ($p > 0.05$), as a possible consequence of the RH% rise observed from May 2018 (Figure 3). Surprisingly, the rind part of the cheese did not show a significant effect of ripening time ($p > 0.05$).

At 12 months of ripening time, TCW showed a slightly higher moisture content than TNCW; however, this difference was not significant in the different cheese zones, with the exception of the inner part of the cheese. At 16 months of storage, this slight difference in moisture was equilibrated, because of the previously described decrease of moisture of TCW cheeses.

Water activity (a_w) followed a decreasing trend from 12 to 16 months of ripening time for both TCW and TNCW cheeses. The decrease of this parameter was significant ($p < 0.05$) for all the different cheese zones. In particular, it should be noted that the decrease of a_w for TNCW cheeses was not related to a decrease in the moisture content. This decrease could be then mainly due to the breakage of proteins in peptides during ripening and the consequent formation of ionic groups that can bind water [29]. The values and distribution of a_w among zones were comparable to those measured in Trentin Grana cheeses ripened for 9 and 18 months [30] and in Gouda [28] and slightly higher than those reported for Asiago d'Allevo [23].

The underrind and the inner zone of the cheese showed a significant main effect of the warehouse for a_w ($p < 0.05$); in particular, a_w was slightly lower in the case of TNCW than TCW. Without other experimental variables, this difference can be explained by the different temperature of ripening of the two warehouses (Figure 3). A higher ripening temperature in the case of TNCW may be responsible for a higher rate of proteolysis in the cheese matrix [31] that may cause an improvement in the water binding capacity because of the increase of hydrophilic interactions caused by the release of peptides [32].

For both moisture content and a_w, the blocking factor, represented by the different cheese producers, did not show a significant main effect in none of the cheese zones ($p > 0.05$, results not shown); consequently, the natural starters used by producers 2 and 4 did not show an impact on these cheese parameters.

3.3. Textural Changes of the Cheese

The changes in textural parameters of Nostrano Valtrompia cheeses are reported in Figure 4a–c. The textural parameters considered (Young's modulus, strain and stress at fracture) followed the opposite trend of the moisture content reported for the different zones of the cheese (Table 2) in accordance with [16]; accordingly, moisture content was found to be negatively correlated with textural parameters (−0.838, −0.817, −0.754 for Young's modulus, stress and strain at rupture, respectively).

In particular, Young's modulus has been demonstrated to have a good correlation with sensory firmness, in the case of hard cheeses [33]. This parameter in the inner part of the cheese, which ranged between ~1.5 and ~3.5 MPa, was found to be higher than literature data of Parmigiano-Reggiano cheese (~1.3 MPa at 12 months of ripening) [34]. This value was also higher than the one reported for other hard and semi-hard cheeses that are usually ripened for a shorter time than Nostrano Valtrompia and some of them are obtained with whole milk, such as Gouda [35,36], Pecorino [37], Manchego [38], Reggianito Argentino [39] and Cheddar [17,40].

The cheese rind was found to be more elastic than, in order, the underrind and the inner part of the cheese, which were found to be more brittle and fragile [16,41].

In general, Young's modulus and stress at fracture slightly increased as a consequence of the increasing ripening time (from 12 to 16 months) in all the measured cheese zones; this trend was in accordance with the one observed in Parmigiano-Reggiano cheese [34]. In particular, the increase of Young's modulus during ripening time was significant ($p < 0.05$) for TNCW cheeses in all the zones; conversely, in the case of TCW cheeses, the increase of this parameter was significant only in the case of the rind zone (Figure 4a). Stress and strain at fracture were significantly affected by ripening time only in the rind part of the cheese ($p < 0.05$) (Figure 4b,c); in particular, strain at fracture showed a significant decrease from 12 to 16 months of storage for both TNCW and TCW cheeses, indicating the formation of a more brittle rind, despite the presence of linseed oil on the cheese surface.

By comparing the two different ripening conditions, it was possible to highlight some slight differences; in general, Young's modulus and stress at facture were slightly lower in the case of TCW than in the case of TNCW in all the three cheese zones.

The producer of the cheese (the blocking factor of the design) showed a significant main effect for all the evaluated textural parameters in all the three cheese zones ($p < 0.05$, results not shown), with the exception of rupture force and Young's modulus in the underrind. Cheeses from producer 4 showed the highest stress, strain at fracture and Young's modulus (5.6 ± 1.5 MPa, 2.5 ± 0.5, 4.6 ± 1.6 MPa, respectively), while producer 2 showed the lowest values (3.1 ± 0.5 MPa, 1.5 ± 0.3, 3.8 ± 1.3 MPa). Accordingly, the use of the natural starters did not group producers 2 and 4; therefore, the use of natural straters could not be considered a characterizing point of Nostrano Valtrompia cheesemaking production because of the low dosage and the high bacterial count of the autoctonous microflora that can develop during the early stage of cheesemaking. Differences among producers that can be related to the relatively high variability of cheesemaking steps (Table 1) were intentionally blocked into the statistical model [42,43] in order to more accurately observe differences related to the ripening time and the different warehouses.

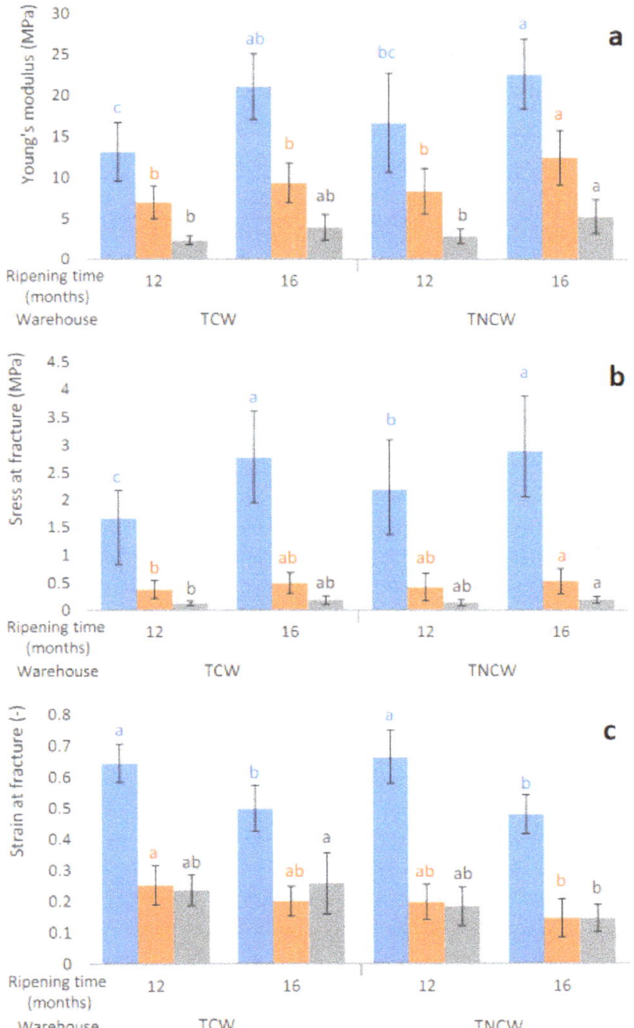

Figure 4. Young's modulus (**a**), stress (**b**) and strain at fracture (**c**) of Nostrano Valtrompia cheeses ripened for 12 and 16 months in temperature conditioned warehouse (TCW) and traditional non conditioned warehouse (TNCW). Blue: rind; orange: underrind; grey: inner part of the cheese. Lower case letters (a–c) indicate significant differences ($p < 0.05$) within the same cheese zone (rind, underrind, inner part).

3.4. Color Characteristics of the Cheese

Cheese color is an important quality parameter that can be related to the cheese ripening time and can affect consumer's appreciation [44,45].

As it is possible to observe from Table 3, b^* values did not show a significant change ($p > 0.05$) in relation with both the ripening time and the warehouses considered in this study; on the contrary, L^* value showed a significant increase ($p < 0.05$) between 12 and 16 months of ripening time in the case of TNCW cheeses in the inner and in the underrind zone. Moreover, the a^* value showed a significant but slight increase during ripening ($p < 0.05$) in the case of TNCW cheeses in the underrind zone.

Table 3. Color parameters of Nostrano Valtrompia cheeses ripened for 12 and 16 months in temperature conditioned warehouse (TCW) and traditional non conditioned warehouse (TNCW).

Color Parameter	Warehouse	Ripening Time (Months)	Rind	Cheese Zone Underrind	Inner Part
L^*	TCW	12	50.7 ± 4.8	58.3 [b] ± 2.8	66.1 [b] ± 5.8
		16	48.6 ± 1.9	59.2 [ab] ± 2.5	67.5 [ab] ± 4.1
	TNCW	12	49.3 ± 3.9	58.3 [b] ± 3.5	65.8 [b] ± 6.1
		16	48.8 ± 5.4	61.0 [a] ± 3.1	70.2 [a] ± 2.9
a^*	TCW	12	8.0 ± 1.1	−1.0 [ab] ± 0.9	−0.7 ± 1.0
		16	8.6 ± 1.2	−0.8 [ab] ± 0.8	−0.6 ± 0.9
	TNCW	12	7.0 ± 2.0	−1.2 [b] ± 0.8	−1.2 ± 0.8
		16	6.1 ± 1.9	−0.6 [a] ± 0.8	−0.8 ± 1.0
b^*	TCW	12	17.7 ± 5.6	20.4 ± 2.8	22.7 ± 2.7
		16	16.7 ± 3.3	20.2 ± 3.1	23.5 ± 3.6
	TNCW	12	15.1 ± 6.3	19.5 ± 2.0	22.2 ± 2.9
		16	11.9 ± 3.2	21.1 ± 3.0	24.0 ± 3.0

lowercase letters indicate significant differences ($p < 0.05$) within the same column.

An increase in the lightness index L^* during ripening time is an interesting finding for a long-ripened cheese type, as previous studies performed on various ripened cheeses reported a decrease of this parameter during ripening [23,24,37,46,47], that can be due to an increase in the hydration of proteins and a reduction of light scattering related to free water [48]. On the other hand, Sberveglieri et al. [49] recently reported an increase in the L^* index of Parmigiano Reggiano cheese during a 36-month period.

L^* index was greater in the inner part than in the underrind and in the rind part of the cheese; this was probably due to the presence of a moisture gradient in the cheese zones, as light scattering phenomena can be inversely related to the moisture content [50], and because of the presence of linseed oil in the rind, that showed the lowest L^* values.

Cheeses were characterized by a strongly yellow color (high b^* values) that can also be related to the use of saffron and that is in line with the values reported for other cheeses manufactured with the addition of saffron [10,11]; because of the stability of crocetin esters from oxidation, during storage time b^* value was not lowered [11].

In general, L^* values were in line [51] or lower [47,49] with those reported for Parmigiano Reggiano cheese ripened up to 50 months; by considering other PDO ripened Italian cheeses such as Montasio and Asiago cheeses, L^* a^* b^* values were comparable, with the exception of b^* values in the case of Asiago, which were lower [23,24,52].

Concerning the different producers, the blocking factor showed significant main effects ($p < 0.05$, results not shown) for a^* and b^* in the underrind and inner part of the cheeses. Differences in color characteristic of the inner part of the cheeses may be caused by the different quantities of saffron used by the producers.

3.5. Sensory Properties of the Cheeses

Discriminant triangle test showed that sensorial differences in TCW and TNCW cheeses were successfully detected in 62% and 63% of the cases, for cheeses ripened for 12 and 16 months, respectively. Therefore, TCW and TNCW were significantly sensorially different at both 12 and 16 months of storage ($p < 0.001$).

The descriptive sensory analysis performed by local consumers of Nostrano Valtrompia cheese, whose results are reported in Figure 5, allowed to observe different evaluations related to both the ripening warehouse and time. In particular, TNCW cheeses at 16 months of ripening time were characterized by a differently ($p < 0.05$) perceived presence of openings and eyes than TCW cheeses and TNCW at 12 months of ripening. This result is in accordance with the higher apparent porosity of TNCW cheeses, measured with image analysis (Section 3.6); as previously stated, it can be related to the different temperature environmental conditions of this warehouse, which showed larger daily and seasonal variations and higher maximum temperatures than TCW [53]. Concerning the overall appreciation, TCW cheeses ripened for 12 months showed slightly lower but not significantly differently scores ($p > 0.05$) than TNCW cheeses and TCW cheeses at 16 months of storage. Similarly, aroma and taste scores were slightly lower for TCW cheeses when compared with TNCW cheeses; again, this difference was not significant, despite being at the limit of significance ($p = 0.07$). No significant differences were also detected for firmness and color ($p > 0.05$).

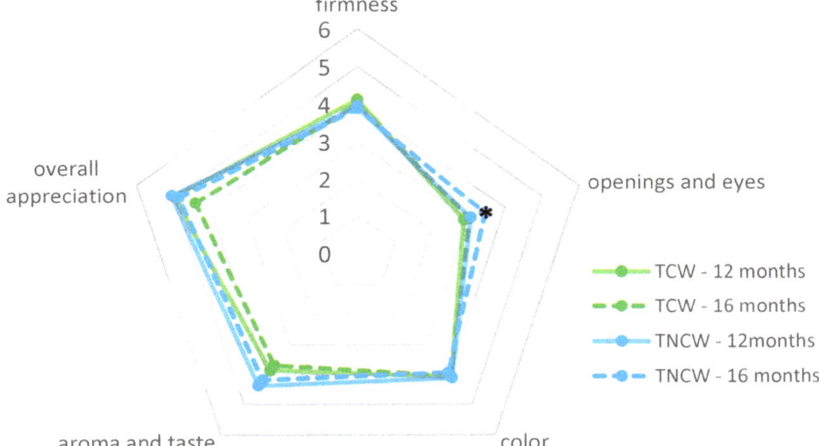

Figure 5. Quantitative descriptive analysis scores (1–7 ordinal scale) for sensory attributes evaluated by Nostrano Valtrompia cheese consumers. The mean score value (4) represents the optimal value for firmness, aroma and taste, color, opening and eyes; for overall appreciation the maximum value of 7 represents the ideal best score. Asterisks (*) represent a significant difference ($p < 0.05$) between treatments for the specific sensory attribute.

Cheese color and the presence of openings and eyes showed the main effect of the blocking factor (producer) ($p < 0.05$, results not shown); on the contrary, the other sensory parameters did not determine a significant main effect of the producer. In particular, for cheese openings and eyes, producer 4 showed the highest score (3.5) while producer 2 the lowest (2.5); conversely, for color, producer 3 (4.7) showed the highest score, while producers 1, 2 and 4 determined similar results (3.8, 3.9 and 3.9, respectively).

3.6. Cheese Openings

The presence of openings in the cheese macrostructure can be a typical feature for certain cheese varieties (e.g., eyes in the Swiss-type) or considered defects when a uniform, close texture is wanted [54]. In the case of Nostrano Valtrompia, uniformly distributed medium fine eyes are expected and accepted [1–3]. Figure 6a–d show the paste appearance of Nostrano Valtrompia ripened in TCW and TNCW, for 12 and 16 months. The use of image analysis allowed the definition of size (mm^2) and shape of the openings.

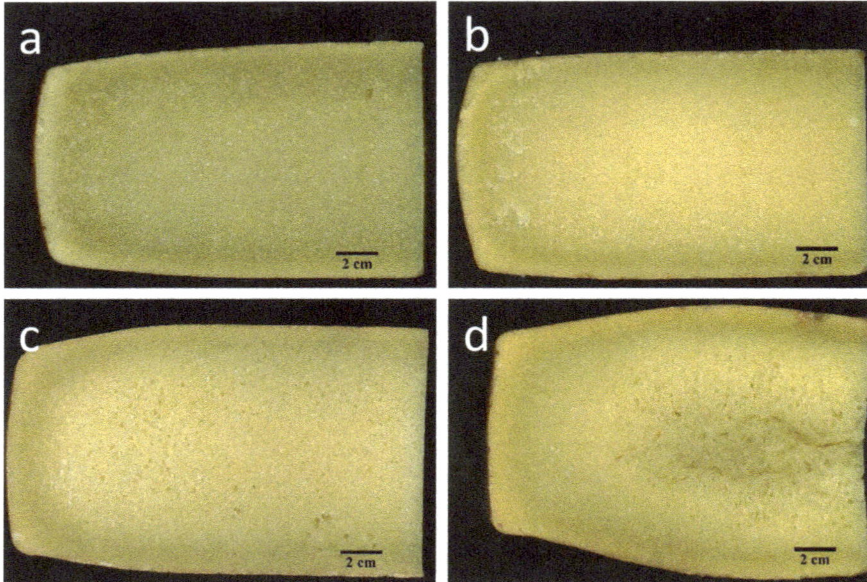

Figure 6. Images of Nostrano Valtrompia cheeses ripened for 12 and 16 months in temperature conditioned warehouse (TCW) and traditional non conditioned warehouse (TNCW). (**a**) TCW at 12 months of ripening; (**b**) TCW at 16 months of ripening; (**c**) TNCW at 12 months of ripening; (**d**) TNCW at 16 months of ripening.

In general, the porosity of Nostrano Valtrompia cheese due to eyes was higher in the case of TNCW (1.17 ± 1.58%) than TCW cheeses (0.56 ± 1.03%), despite not being significantly different ($p > 0.05$). Moreover, the mean porosity due to cracks or splits was higher but not significantly different ($p > 0.05$) for TNCW (1.86 ± 2.32%) than TCW cheeses (0.33 ± 0.38%). The high variability encountered in the porosity of Nostrano Valtrompia cheese can be a consequence of the artisanal and not standardized cheese making procedure (Table 1) performed for this kind of cheese (e.g., commercial/selected starters are not used) and to the microbial diversity of raw milk that could promote different fermentative phenomena [3,55]. Total porosity of Nostrano Valtrompia cheeses ranged between 0.0% and 10.6%, with a mean value of 2.0 ± 2.5%. In particular, the mean value, as expected, was lower than that reported in the case of Swiss-type cheeses (higher than 9.4%) [56] but also lower than that of Montasio cheese ranging from 3.1% to 18.3%) [57]. On the contrary Nostrano Valtrompia cheese porosity was higher than that reported for blue cheeses (~0.10–0.05%) [58] or Pecorino cheese (~0.2%) [37].

Figure S1a–d (Supplementary Material) show a highly distorted, right-skewed size distribution for all the tested ripening times; the most frequent opening size was around ~0.5 mm^2 to ~6 mm^2 for all the treatments. This can also be observed from the highly different mean and median (D50) opening size parameters (Table 4). Other authors reported a similar right-skewed size distribution in the openings of blue cheese [55] and hard/semi-hard Swiss-type cheeses [56,59]. Comparing the two warehouses, it is possible to observe a greater presence of large openings in cheese stored in TNCW; however, this slight variation was not reflected by a significant influence ($p > 0.05$) on the D90 parameter of the two different warehouses (Table 4). Conversely, D50 showed a significant variation between TNCW and TCW cheeses ($p < 0.05$), with the first ones at 16 months of ripening that showed significantly higher values than TCW at 12 and 16 months ($p < 0.05$). The higher temperature reached during the ripening period in TNCW (Figure 3a) could promote the growth of a different cheese microbiota and/or accelerate the fermentation rate. Considering the presence of mesophilic heterofermentative species among the various microorganisms known to contribute to the Nostrano Valtrompia ripening [2,3], a higher temperature could have boosted the production of gaseous metabolites leading to an increase of size and number of openings.

Table 4. Size frequency distribution percentiles (D10, D50, D90), mean and span (Equation (5)) of Nostrano Valtrompia cheese openings measured with image analysis.

Warehouse	Ripening Time (Months)	D10 (mm^2)	D50 (mm^2)	D90 (mm^2)	Span (mm^2)	Mean (mm^2)
TCW	12	1.1 ± 0.4	3.8 [b] ± 3.6	15.1 ± 15.6	3.7 ± 2.3	6.8 ± 5.7
	16	1.0 ± 0.4	3.0 [b] ± 2.7	13.0 ± 10.6	3.6 ± 2.7	6.3 ± 4.2
TNCW	12	1.0 ± 0.3	4.5 [ab] ± 2.1	42.7 ± 48.2	9.3 ± 12.3	16.6 ± 13.9
	16	2.6 ± 2.7	6.4 [a] ± 4.0	35.5 ± 27.9	7.3 ± 7.1	26.7 ± 33.1

lowercase letters indicate significant differences ($p < 0.05$) within the same column.

The presence of some cracks, mainly in TNCW cheeses, may be favoured by the low moisture content and by a poor plasticity of the casein network that sometimes locally can originate the crack when the pressure of gas deriving from fermentation is not able to create an eye and fractures the paste and/or by the inability of the cheese to expand its volume when the temperature of the warehouse increases [60].

4. Conclusions

Ripening of Nostrano Valtrompia PDO cheeses in a temperature conditioned warehouse (TCW) and in a traditional not conditioned warehouse (TNCW) allowed for the production of cheeses characterized by some different and peculiar traits but that in general were pooled by a common identity. TCW created less variable environmental ripening conditions than TNCW; TCW cheeses were characterized by a slightly softer texture, a slightly different porosity distribution and a different sensory perception than TNCW cheeses. Despite these differences, the overall appreciation of the cheese, as well as other physico-chemical and sensory traits, was not affected by the different ripening conditions.

The results of this study suggest that it is possible to consider different types of warehouses for Nostrano Valtrompia ripening, in order to increase the overall capacity of the Nostrano Valtrompia PDO system to store cheeses until the end of the 12-month ripening period established by the PDO regulation. At present, the high number of cheeses produced in the PDO area exceed the storage capacity. As a consequence, part of these cheeses is ripened in other areas out of PDO borders, and it is excluded from the Nostrano Valtrompia PDO system. The studied relations between cheese storage conditions and quality is the base to evaluate the possibility of introducing new warehouses in the area. These results can be helpful to introduce in the area new warehouses, such as an old mine gallery that could be useful to guarantee a relatively constant temperature and to reduce the need for environmental controls.

Supplementary Materials: The following are available online at http://www.mdpi.com/2304-8158/9/8/1101/s1, Figure S1: Frequency distributions (%) of Nostrano Valtrompia openings measured with image analysis. Nostrano Valtrompia cheeses were ripened for 12 and 16 months in temperature conditioned warehouse (TCW) and traditional non conditioned warehouse (TNCW). (a) TNCW at 12 months of ripening; (b) TCW at 12 months of ripening; (c) TNCW at 16 months of ripening; (d) TCW at 16 months of ripening.

Author Contributions: Conceptualization, G.M., R.M.; methodology, L.B. and M.A.; software, M.A. and L.B.; validation, L.B. and M.A.; formal analysis, L.B.; investigation, L.B.; resources, G.M. and R.M.; data curation, L.B. and M.A.; writing—original draft preparation, M.A. and G.M.; writing—review and editing, L.B.; visualization, M.A. and L.B.; supervision, G.M.; project administration, G.M. and R.M.; funding acquisition, G.M. All authors have read and agreed to the published version of the manuscript.

Funding: This research was funded by Consorzio Tutela Nostrano Valtrompia DOP, in the frame of the project "Progetto Pilota per la Stagionatura del Formaggio DOP Nostrano Valtrompia in Miniera" grant number 2016/00348318 from Giunta Regionale della Lombardia n. 11567 del 18 dicembre 2015 «Programma di Sviluppo Rurale 2014–2020 della Lombardia. Operazione 16.2.01 Progetti pilota e sviluppo dell'innovazione" and "The APC was funded by Food and Drug Department, University of Parma".

Conflicts of Interest: The authors declare no conflicts of interest.

References

1. EC Commission Implementing Regulation (EU) N° 629/2012 of 6 July 2012 entering a name in the register of protected designations of origin and protected geographical indications Nostrano Valtrompia (PDO). *Off. J. Eur. Union* **2012**, *L182*, 12.
2. Mucchetti, G.; Neviani, E. *Microbiologia e Tecnologia Lattiero-Casearia*; Tecniche nuove: Milan, Italy, 2006; ISBN 8848168175.
3. Gobbetti, M.; Neviani, E.; Fox, P. *The Cheeses of Italy: Science and Technology*; Springer International Publishing AG: Cham, Switzerland, 2018; ISBN 9783319898544.
4. Mucchetti, G.; Taglietti, P.; Emaldi, G.C.; Mondinelli, R. Influenza di alcuni fattori tecnologici ed ambientali sulle caratteristiche organolettiche del formaggio di Valle Trompia. *Latte* **1993**, *18*, 548–556.
5. Mucchetti, G.; Neviani, E.; Carminati, D.; Contarini, G.; Olivari, G.; Mondinelli, R.; Savino, L. Il formaggio "Nostrano" di Valle Trompia: Indagine tecnologica e composizione chimica. *Ind. Latte* **1985**, *21*, 23–40.
6. Mucchetti, G. Tecnologia di caseificazione del formaggio Bagoss. *Sci. Tec. Latt. Casearia* **1996**, *47*, 125–135.
7. Addeo, F.; Mucchetti, G. Produzioni casearie tipiche: Aspetti qualitativi. *Latte* **2001**, *26*, 56–72.
8. Zannoni, M. Evolution of the sensory characteristics of Parmigiano-Reggiano cheese to the present day. *Food Qual. Prefer.* **2010**, *21*, 901–905. [CrossRef]
9. Carpino, S.; Rapisarda, T.; Belvedere, G.; Licitra, G. Volatile fingerprint of Piacentinu cheese produced with different tools and type of saffron. *Small Rumin. Res.* **2008**, *79*, 16–21. [CrossRef]
10. Licón, C.C.; Carmona, M.; Molina, A.; Berruga, M.I. Chemical, microbiological, textural, color, and sensory characteristics of pressed ewe milk cheeses with saffron (*Crocus sativus* L.) during ripening. *J. Dairy Sci.* **2012**, *95*, 4263–4274. [CrossRef]
11. Aktypis, A.; Christodoulou, E.D.; Manolopoulou, E.; Georgala, A.; Daferera, D.; Polysiou, M. Fresh ovine cheese supplemented with saffron (*Crocus sativus* L.): Impact on microbiological, physicochemical, antioxidant, color and sensory characteristics during storage. *Small Rumin. Res.* **2018**, *167*, 32–38. [CrossRef]
12. Lebecque, A.; Laguet, A.; Devaux, M.F.; Dufour, É. Delineation of the texture of Salers cheese by sensory analysis and physical methods. *Lait* **2001**, *81*, 609–623. [CrossRef]
13. Costa, M.J.; Maciel, L.C.; Teixeira, J.A.; Vicente, A.A.; Cerqueira, M.A. Use of edible films and coatings in cheese preservation: Opportunities and challenges. *Food Res. Int.* **2018**, *107*, 84–92. [CrossRef] [PubMed]
14. Juita, J.; Dlugogorski, B.Z.; Kennedy, E.M.; Mackie, J.C. Low temperature oxidation of linseed oil: A review. *Fire Sci. Rev.* **2012**, *1*, 3. [CrossRef]
15. AOAC. *Official Methods of Analysis of the AOAC. Volume 2*, 15th ed.; Association of Official Analytical Chemists, Inc.: Rockville, MD, USA, 1990; Volume 2, ISBN 0-935584-42-0.
16. Alinovi, M.; Mucchetti, G.; Tidona, F. Application of NIR spectroscopy and image analysis for the characterisation of grated Parmigiano-Reggiano cheese. *Int. Dairy J.* **2019**, *92*, 50–58. [CrossRef]
17. Hort, J.; Le Grys, G. Developments in the textural and rheological properties of UK Cheddar cheese during ripening. *Int. Dairy J.* **2001**, *11*, 475–481. [CrossRef]
18. Hort, J.; Grys, G.; Woodman, J. The relationships between the chemical, rheological and textural properties of cheddar cheese. *Lait* **1997**, *77*, 587–600. [CrossRef]
19. Caccamo, M.; Melilli, C.; Barbano, D.M.; Portelli, G.; Marino, G.; Licitra, G. Measurement of gas holes and mechanical openness in cheese by image analysis. *J. Dairy Sci.* **2004**, *87*, 739–748. [CrossRef]
20. Alinovi, M.; Cordioli, M.; Francolino, S.; Locci, F.; Ghiglietti, R.; Monti, L.; Tidona, F.; Mucchetti, G.; Giraffa, G. Effect of fermentation-produced camel chymosin on quality of Crescenza cheese. *Int. Dairy J.* **2018**, *84*, 72–78. [CrossRef]
21. Alinovi, M.; Rinaldi, M.; Mucchetti, G. Spatiotemporal characterization of texture of Crescenza cheese, a soft fresh Italian cheese. *J. Food Qual.* **2018**, *2018*, 1–8. [CrossRef]
22. Guidetti, R.; Mora, R.; Zannoni, M. Effect of storage conditions on Parmigiano-Reggiano cheese. II. Cheese stored with and without air conditioning. *Sci. e Tec. Latt. Casearia* **1995**, *46*, 178.
23. Marchesini, G.; Balzan, S.; Segato, S.; Novelli, E.; Andrighetto, I. Colour traits in the evaluation of the ripening period of Asiago cheese. *Ital. J. Anim. Sci.* **2009**, *8*, 412–413. [CrossRef]

24. Aprea, E.; Romanzin, A.; Corazzin, M.; Favotto, S.; Betta, E.; Gasperi, F.; Bovolenta, S. Effects of grazing cow diet on volatile compounds as well as physicochemical and sensory characteristics of 12-month-ripened Montasio cheese. *J. Dairy Sci.* **2016**, *99*, 6180–6190. [CrossRef] [PubMed]

25. Malacarne, M.; Summer, A.; Franceschi, P.; Formaggioni, P.; Pecorari, M.; Panari, G.; Mariani, P. Free fatty acid profile of Parmigiano-Reggiano cheese throughout ripening: Comparison between the inner and outer regions of the wheel. *Int. Dairy J.* **2009**, *19*, 637–641. [CrossRef]

26. Innocente, N.; Munari, M.; Biasutti, M. Characterization by solid-phase microextraction-gas chromatography of the volatile profile of protected designation of origin Montasio cheese during ripening. *J. Dairy Sci.* **2013**, *96*, 26–32. [CrossRef] [PubMed]

27. Gatti, M.; Lindner, J.D.; Gardini, F.; Mucchetti, G.; Bevacqua, D.; Fornasari, M.E.; Neviani, E. A model to assess lactic acid bacteria aminopeptidase activities in Parmigiano Reggiano cheese during ripening. *J. Dairy Sci.* **2008**, *91*, 4129–4137. [CrossRef]

28. Salazar, J.K.; Carstens, C.K.; Ramachandran, P.; Shazer, A.G.; Narula, S.S.; Reed, E.; Ottesen, A.; Schill, K.M. Metagenomics of pasteurized and unpasteurized gouda cheese using targeted 16S rDNA sequencing. *BMC Microbiol.* **2018**, *18*, 1–13. [CrossRef]

29. Ak, M.M.; Gunasekaran, S. Dynamic Rheological Properties of Mozzarella Cheese During Refrigerated Storage. *J. Food Sci.* **2006**, *61*, 566–569. [CrossRef]

30. Cavazza, A.; Franciosi, E.; Poznanski, E.; Monfredini, L.; Settanni, L. The spatial distribution of bacteria in Grana-cheese during ripening. *Syst. Appl. Microbiol.* **2012**, *35*, 54–63. [CrossRef]

31. Folkertsma, B.; Fox, P.F.; McSweeney, L.H. Accelerated Ripening of Cheddar Cheese at Elevated Temperatures. *Int. Dairy J.* **1996**, *6*, 1117–1134. [CrossRef]

32. Sforza, S.; Cavatorta, V.; Lambertini, F.; Marchelli, R.; Dossena, A.; Galaverna, G. Cheese peptidomics: A detailed study on the evolution of the oligopeptide fraction in Parmigiano-Reggiano cheese from curd to 24 months of aging. *J. Dairy Sci.* **2012**, *95*, 3514–3526. [CrossRef]

33. Ibanescu, L.; Allard, T.; Fonseca, F.; Caroline, P.; Buchin, S.; Salles, C.; Dibie, J.; Guichard, E. Relating transformation process, eco-design, composition and sensory quality in cheeses using PO 2 ontology. *Int. Dairy J.* **2019**, *92*, 1–10. [CrossRef]

34. Noël, Y.; Zannoni, M.; Hunter, E.A. Texture of Parmigiano Reggiano cheese: Statistical relationships between rheological and sensory variates. *Lait* **1996**, *76*, 243–254. [CrossRef]

35. Vandenberghe, E.; Choucharina, S.; De Ketelaere, B.; De Baerdemaeker, J.; Claes, J. Spatial variability in fundamental material parameters of Gouda cheese. *J. Food Eng.* **2014**, *131*, 50–57. [CrossRef]

36. Vandenberghe, E.; Charalambides, M.N.; Mohammed, I.K.; De Ketelaere, B.; De Baerdemaeker, J.; Claes, J. Determination of a critical stress and distance criterion for crack propagation in cutting models of cheese. *J. Food Eng.* **2017**, *208*, 1–10. [CrossRef]

37. Rinaldi, M.; Chiavaro, E.; Massini, R. Pecorino of appennino reggiano cheese: Evaluation of ripening time using selected physical properties. *Ital. J. Food Sci.* **2010**, *22*, 54–59.

38. Martín-Alvarez, P.J.; Cabezas, L. Relationship between sensory and instrimental measurements of texture for artisanal and industrial Manchego cheeses. *J. Sens. Stud.* **2007**, *22*, 462–476. [CrossRef]

39. Bertola, N.C.; Bevilacqua, A.E.; Zaritzky, N.E. Rheological Behaviour of Reggianito Argentino Cheese Packaged in Plastic Film During Ripening. *LWT-Food Sci. Technol.* **1995**, *615*, 610–615. [CrossRef]

40. Charalambides, M.N.; Williams, J.G.; Chakrabarti, S. A study of the influence of ageing on the mechanical properties of Cheddar cheese. *J. Mater. Sci.* **1995**, *30*, 3959–3967. [CrossRef]

41. Patel, H.G.; Upadhyay, K.G.; Miyani, R.V.; Pandya, A.J. Instron texture profile of buffalo milk cheddar cheese as influenced by composition and ripening changes. *Food Qual. Prefer.* **1993**, *4*, 187–192. [CrossRef]

42. Tsai, P.W. A study of two types of split-plot designs. *J. Qual. Technol.* **2016**, *48*, 44–53. [CrossRef]

43. Alinovi, M.; Wiking, L.; Corredig, M.; Mucchetti, G. Effect of frozen and refrigerated storage on proteolysis and physico-chemical properties of high-moisture citric Mozzarella cheese. *J. Dairy Sci.* **2020**, *103*. [CrossRef]

44. Wadhwani, R.; McMahon, D.J. Color of low-fat cheese influences flavor perception and consumer liking. *J. Dairy Sci.* **2012**, *95*, 2336–2346. [CrossRef] [PubMed]

45. Buffa, M.N.; Trujillo, A.J.; Pavia, M.; Guamis, B. Changes in textural, microstructural, and colour characteristics during ripening of cheeses made from raw, pasteurized or high-pressure-treated goats' milk. *Int. Dairy J.* **2001**, *11*, 927–934. [CrossRef]

46. Romani, S.; Sacchetti, G.; Pittia, P.; Pinnavaia, G.G.; Dalla Rosa, M. Physical, Chemical, Textural and Sensorial Changes of Portioned Parmigiano Reggiano Cheese Packed under Different Conditions. *Food Sci. Technol. Int.* **2002**, *8*, 203–211. [CrossRef]

47. D'Incecco, P.; Limbo, S.; Hogenboom, J.; Rosi, V.; Gobbi, S.; Pellegrino, L. Impact of extending hard-cheese ripening: A multiparameter characterization of Parmigiano reggiano cheese ripened up to 50 months. *Foods* **2020**, *9*, 268. [CrossRef]

48. Sánchez-Macías, D.; Fresno, M.; Moreno-Indias, I.; Castro, N.; Morales-delaNuez, A.; Alvarez, S.; Argüello, A. Physicochemical analysis of full-fat, reduced-fat, and low-fat artisan-style goat cheese. *J. Dairy Sci.* **2010**, *93*, 3950–3956. [CrossRef]

49. Sberveglieri, V.; Bhandari, M.P.; Carmona, E.N.; Betto, G.; Soprani, M.; Malla, R.; Sberveglieri, G. Spectrocolorimetry and nanowire gas sensor device S3 for the analysis of Parmigiano Reggiano cheese ripening. In Proceedings of the ISOCS/IEEE International Symposium on Olfaction and Electronic Nose, Montreal, QC, Canada, 28–31 May 2017; pp. 9–11.

50. Alinovi, M.; Mucchetti, G. Effect of freezing and thawing processes on high-moisture Mozzarella cheese rheological and physical properties. *LWT-Food Sci. Technol.* **2020**, *124*, 109137. [CrossRef]

51. Severini, C.; Bressa, F.; Romani, S.; Dalla Rosa, M. Physical and chemical changes in vacuum packaged Parmigiano Reggiano cheese during storage at 25, 2 and -25C. *J. Food Qual.* **1998**, *21*, 355–367. [CrossRef]

52. Cozzi, G.; Ferlito, J.; Pasini, G.; Contiero, B.; Gottardo, F. Application of near-infrared spectroscopy as an alternative to chemical and color analysis to discriminate the production chains of Asiago d'Allevo cheese. *J. Agric. Food Chem.* **2009**, *57*, 11449–11454. [CrossRef]

53. Lee, K.; Uegaki, K.; Nishii, C.; Nakamura, T.; Kubota, A.; Hirai, T.; Yamada, K. Computed tomographic evaluation of gas hole formation and structural quality in Gouda-type cheese. *Int. J. Dairy Technol.* **2012**, *65*, 232–236. [CrossRef]

54. Martley, F.G.; Crow, V.L. Open texture in cheese: The contributions of gas production by microorganisms and cheese manufacturing practices. *J. Dairy Res.* **1996**, *63*, 489–507. [CrossRef]

55. Gatti, M.; Bottari, B.; Lazzi, C.; Neviani, E.; Mucchetti, G. Invited review: Microbial evolution in raw-milk, long-ripened cheeses produced using undefined natural whey starters. *J. Dairy Sci.* **2014**, *97*, 573–591. [CrossRef]

56. Schuetz, P.; Guggisberg, D.; Jerjen, I.; Fröhlich-Wyder, M.T.; Hofmann, J.; Wechsler, D.; Flisch, A.; Bisig, W.; Sennhauser, U.; Bachmann, H.P. Quantitative comparison of the eye formation in cheese using radiography and computed tomography data. *Int. Dairy J.* **2013**, *31*, 150–155. [CrossRef]

57. Innocente, N.; Corradini, C. Use of an immage analysis technique in the quality control of Montasio cheese. *Sci. Tec. Latt. Casearia* **1998**, *42*, 82–94.

58. Kebary, K.M.K.; Morris, H.A. Porosity, Specific Gravity and Fat Dispersion in Blue Cheeses. *Food Microstruct.* **1988**, *7*, 153–160.

59. Guggisberg, D.; Fröhlich-Wyder, M.T.; Irmler, S.; Greco, M.; Wechsler, D.; Schuetz, P. Eye formation in semi-hard cheese: X-ray computed tomography as a non-invasive tool for assessing the influence of adjunct lactic acid bacteria. *Dairy Sci. Technol.* **2013**, *93*, 135–149. [CrossRef]

60. White, S.R.; Broadbent, J.R.; Oberg, C.J.; McMahon, D.J. Effect of Lactobacillus helveticus and Propionibacterium freudenrichii ssp. shermanii combinations on propensity for split defect in Swiss cheese. *J. Dairy Sci.* **2003**, *86*, 719–727. [CrossRef]

© 2020 by the authors. Licensee MDPI, Basel, Switzerland. This article is an open access article distributed under the terms and conditions of the Creative Commons Attribution (CC BY) license (http://creativecommons.org/licenses/by/4.0/).

 *foods*

Article

Dairy by-Products Concentrated by Ultrafiltration Used as Ingredients in the Production of Reduced Fat Washed Curd Cheese

Ana Raquel Borges [1], **Arona Figueiroa Pires** [1], **Natalí Garcia Marnotes** [1], **David Gama Gomes** [1], **Marta Fernandes Henriques** [1,2] and **Carlos Dias Pereira** [1,2,*]

[1] Polytechnic Institute of Coimbra, College of Agriculture, Bencanta, 3045-601 Coimbra, Portugal; raquelborges81@gmail.com (A.R.B.); arona@esac.pt (A.F.P.); natali@esac.pt (N.G.M.); david@esac.pt (D.G.G.); mhenriques@esac.pt (M.F.H.)

[2] Research Centre for Natural Resources, Environment and Society (CERNAS), Bencanta, 3045-601 Coimbra, Portugal

* Correspondence: cpereira@esac.pt; Tel.: +351-965411989

Received: 2 July 2020; Accepted: 27 July 2020; Published: 30 July 2020

Abstract: In the following study, three different dairy by-products, previously concentrated by ultrafiltration (UF), were used as ingredients in the production of reduced-fat (RF) washed curd cheeses in order to improve their characteristics. Conventional full-fat (FF) cheeses (45% fat, dry basis (db)) and RF cheeses (20–30% fat, db) were compared to RF cheeses produced with the incorporation of 5% concentrated whey (RF + CW), buttermilk (RF + CB) or sheep second cheese whey (RF + CS). Protein-to-fat ratios were lower than 1 in the FF cheeses, while RF cheeses ranged from 1.8 to 2.8. The tested by-products performed differently when added to the milk used for cheese production. The FF cheese showed a more pronounced yellow colour after 60 and 90 days of ripening, indicating that fat plays an important role regarding this parameter. As far as the texture parameters are concerned, after 60 days of ripening, RF cheeses with buttermilk presented similar results to FF cheeses for hardness (5.0–7.5 N) and chewiness (*ca.* 400). These were lower than the ones recorded for RF cheeses with added UF concentrated whey (RF + CW) and second cheese whey (RF + CS), which presented lower adhesiveness values. RF cheeses with 5% incorporation of buttermilk concentrated by UF presented the best results concerning both texture and sensory evaluation.

Keywords: whey; buttermilk; second cheese whey; ultrafiltration; reduced-fat cheese

---

1. Introduction

The firm texture observed in reduced-fat (RF) and low-fat (LF) cheeses is one of the major problems resulting from fat reduction [1]. Cheese structure can be described as a continuous protein network interrupted by dispersed fat globules, which originate weak points in the protein network. In RF/LF cheeses the para-casein network becomes denser, originating the development of a firm and rubbery texture that does not break down during mastication [2,3].

Fat also plays an important role in the development of cheese flavour and appearance. The loss of flavour in RF/LF cheeses results from the lack of precursors from the fat, the lack of fat as a solvent for flavouring compounds, or the differences in the physical structure of RF/LF cheeses that inhibit certain enzymatic reactions which are essential for the formation of flavouring compounds [4]. Moreover, the observed differences between FF and RF Cheddar cheeses are not solely owed to differences in the cheese matrix and flavour release, but also to differences in ripening biochemistry, which lead to an imbalance of many flavour-contributing compounds [5].

Despite the significant advances in understanding both the biochemical and physicochemical characteristics of RF and LF cheeses and the introduction of technological developments, it is still necessary to evaluate solutions with the potential to improve the flavour, texture and sensory properties of such cheeses.

Strategies to Improve the Characteristics of RF/LF Cheeses

The general approaches that have been used to improve the texture of RF or LF cheese involve decreasing the protein concentration, stimulating proteolysis, or creating a bigger filler phase to limit the density of the para-casein network [6]. These strategies can be divided into three categories: (i) manipulation of process parameters to enhance the moisture level; (ii) selection of specific starter cultures and use of adjunct cultures (i.e., non-starter lactic acid bacteria); and (iii) the use of stabilizers and fat mimetics to improve cheese texture [7].

The process parameters can be modified in order to increase water retention in the curd, which influences texture properties. This can be achieved by using lower coagulation temperatures; increasing the curd grain size; lowering curd scalding temperatures; or by increasing the surface area of the fat globules through milk homogenization [4,8].

The use of starter cultures that produce exopolysaccharide (EPS) can also improve the textural characteristics of LF cheeses by changing the microstructure and proteolysis [9] and has the potential to improve cheese flavour [10,11].

Fat replacers, alone or associated to the manipulation of process parameters, are the most promising alternative for improving the sensory properties of RF/LF cheeses. These ingredients are water-soluble compounds used to replace the functional characteristics of fat. They improve texture and cheese yield as well as the sensory and functional properties by binding water and providing a sense of lubricity and creaminess [12]. Polysaccharides and whey proteins (WP) are the most commonly used fat replacers.

Several reports evaluate the chemical and rheological properties of RF cheese with added polysaccharides as fat replacers. Guar and Arabic gums were tested on Iranian white cheese [13]. Waxy maize starch also increased the moisture content and water holding capacity, improving the overall quality of RF cheeses [14]. The partial replacement of milk fat for inulin increased meltability, cohesiveness and viscosity, while decreasing hardness and adhesiveness of acid casein processed cheeses [15]. Carrageenans also improved the textural and rheological properties of LF cheeses by increasing moisture in non-fat solids (MNFS) [16]. Cheeses containing agave fructans were compared to FF and RF samples without fructans and demonstrated the texturing role of the carbohydrates [17]. Konjac glucomanan was also indicated as a potential fat replacer to be used in Mozzarella [18]. The addition of alginate also improved the textural, microstructural and colour properties of LF Cheddar cheeses [19]. Recently waxy rice starch, sodium carboxymethyl cellulose (CMC) and glutamine transaminase were also tested as texturizers and crosslinking agents in Mozzarella [20].

Native WP can be aggregated to obtain colloidal microparticulated whey protein (MWP). MWP is formed by mixing native whey proteins and protein aggregates. These particles can be manufactured in diameters ranging from 0.1 to 100 μm [21,22]. Their ability to enhance creaminess is based on a "ball bearing mechanism" originally reported by Cheftel and Dumay [23] and recently confirmed [24].

Several aspects have to be pondered when using WP as ingredients for cheese production, such as: (i) the importance of WP denaturation, which allows their contribution to the protein matrix of the cheese; (ii) the increased water-holding capacity of cheese curds; (iii) the lower acidification of the cheeses as a result of the higher buffering capacity of WP; (iv) the occurrence of differences in flavour of the modified products, which tend to be more pronounced during ripening [25].

When using WP as a fat replacer in cheese it is recommended a low ratio of native/denatured WP to ensure its function as an inert filler. It has also been ascertained that particle size should be within the range of 1 to 10 μm in order to avoid disturbance of the para-casein network [26]. However, other authors advocate that higher particle sizes (20–100 μm) do not impart negative effects to cheese properties, namely taste, flavour or consistency [27,28].

Whey proteins and MWP are commercially available as powders. However, small and medium-scale cheese plants can concentrate whey by ultrafiltration and, after proper treatments, they can use the liquid whey concentrates in the cheese production. The same methodology can be applied to other dairy by-products, allowing for their in-plant valorisation. Hence, in the present work, we selected three liquid by-products of the dairy industry, namely: buttermilk, whey and sheep second cheese whey (the by-product resulting from the production of whey cheese), all previously concentrated by ultrafiltration (UF), to be tested as ingredients in the production of experimental RF/LF cheeses.

2. Materials and Methods

The three reduced-fat, cow's milk cheeses produced with different dairy by-products as ingredients were compared with conventional full-fat cheese (FF) (>45% fat db) and conventional reduced-fat (or half-fat) cheese (RF) (25–45% fat db) (without any addition). The tested fat replacers were, buttermilk (CB), whey (CW) and sheep's second cheese whey (CS).

Second cheese whey, known as *Sorelho*, is the by-product resulting from the manufacture of *Requeijão*, the Portuguese whey cheese, which is produced by thermal aggregation of whey proteins (at ca. 90 °C for 20 min). This product still contains approximately 50% of the original dry matter of the original whey. Lactose and minerals largely contribute to its dry mass, but residual fat and non-thermally precipitated nitrogen components are still present.

All the by-products were previously concentrated by ultrafiltration (UF) with a volumetric concentration factor of ca. 15 (VCF = VFeed/VRetentate). The five cheese products were coded as: (i) conventional full-fat cheese (FF); (ii) reduced-fat cheese (RF); (iii) reduced-fat cheese with buttermilk (RF + CB); (iv) reduced-fat cheese with whey (RF + CW) and; (v) reduced-fat cheese with sheep second cheese whey (RF + CS).

UF concentrates were obtained by following the procedure described by Henriques and co-workers [25] although with small adjustments, namely the ultrafiltration process temperature (40–45 °C) and the smaller membrane cut-off (10 kDa). In the case of CW, after concentration, the retentate was submitted to thermal denaturation (90 °C for 20 min) prior to homogenization at 10 MPa, in a homogenizer Rannie™ model Blue Top (APV, Albertslund, Denmark) and kept frozen (−25 °C) until the moment of use. Buttermilk and *Sorelho* were concentrated by UF using the same conditions and pasteurized at 75 °C for 5 min and kept frozen at −25 °C prior to their incorporation into the milk.

The bovine milk (Quinta da Cioga, Portugal) was delivered to the dairy pilot plant at *Escola Superior Agrária de Coimbra* (ESAC). Part of the milk was skimmed in a Westfalia™ type ADB centrifuge (Westfalia Separator, Germany) and standardised to 3.5% (*v/v*) of fat (for the production of FF cheese) and to 1.5% (*v/v*) of fat (for the production of RF cheeses). At this stage, according to the batches to be produced, the fat replacers (5% *v/v*) were added. Each batch was made up of 40 L.

Milk pasteurization took place at 74 ± 1 °C for 30 s in Pasilac Therm™ (PHE Nordic, Denmark) plate and frame heat exchanger. After temperature stabilization of the mixtures at 29.5 ± 0.5 °C, 0.2 mL^{-1} CaCl$_2$ solution (51% *w/v*) (supplied by Tecnilac, Portugal), starter culture (Mesófilo Plus Starter, Enzilab, Portugal) (10 mgL^{-1}, containing *Lactococcus lactis* subsp. *lactis*, *Lactococcus lactis* subsp. *cremoris* and *Streptococcus thermophilus*), KNO$_3$ (25 mgL^{-1}) and 20 mgL^{-1} animal rennet (>92 g/100 g chimosin, supplied by Tecnilac, Portugal) were added to the milk formulations and mixed thoroughly. The coagulation of the mixtures was performed for approximately 45 min at 30 ± 1 °C. When coagulation was completed, grids were used to cut the curd into small pieces (2 cm^3), in order to promote whey drainage. After draining half of the whey, the same amount of salted hot water (1% *w/v* salt, 30 °C) was added to the curd and the mixture was thoroughly agitated prior to the final whey drainage. The recovered curd was then placed into plastic moulds before being pressed and stored in a refrigerated chamber at 8–9 °C, for 24 h. After this period, cheeses, weighing approximately 250 g (ca. 4.5 cm height and 8.0 cm diameter) were immersed in a brine solution (18–20 °Baumé) for 1.5 h and finally transferred to the ripening chamber (10 ± 2 °C) being kept there for 90 days.

The chemical composition of cheeses, colour, texture, pH and titratable acidity were assessed on the 1st, 30th, 60th and 90th days of ripening. Each physicochemical parameter was evaluated in triplicate.

Cheese moisture was determined by drying the cheese sample in an oven at 105 °C for 24 h according to AOAC method 248.12 Dried samples were tested for ash content in a muffle furnace at 550 °C for 4 h (AOAC 935.42) [29].

The fat content was determined using the Van Gulik method (ISO 3433, 2008) [30]. Cheese protein was determined by multiplying the total nitrogen content of the samples, obtained using the Kjeldahl procedure (AOAC 920.132) [29], by a factor of 6.38.

The pH was measured directly from the cheeses, using a pH meter (PHM61 Laboratory pH Meter, Denmark) equipped with a probe for reading solids and the titratable acidity was expressed as g of lactic acid/100 g cheese (AOAC 920.124) [29].

According to the physicochemical composition of each cheese sample, moisture in non-fat solids (MNFS) and fat in dry matter (FDM) were calculated.

Colour was expressed by the individual three coordinates of CIE L*(lightness), a*(red-green axis) b*(blue-yellow axis) system using a Chroma Minolta CR-200B colorimeter (Japan). For each cheese type, three readings for colour were performed on the rind and on the paste of two cheeses ($n = 6$).

A Stable Micro Systems Texture analyzer, model TA.XT Express Enhanced (Stable Micro Systems, Surrey, UK), was used to perform textural analysis and the results were calculated by the Specific Expression PC Software. A texture profile analysis was run with a penetration distance of 15 mm at 1 mm/s test speed, using an acrylic cylindrical probe with a diameter of 12.5 mm and height of 38.1 mm. The following parameters were quantified: hardness (N) (the peak force measured during the first compression cycle), adhesiveness (g.s) (the negative force area for the first bite, representing the necessary work to pull the compression plunger away from the sample), cohesiveness (the ratio between the positive area during the second compression and the area during the first compression) and chewiness (the product of gumminess and springiness) were quantified. Three penetrations were performed on the surface (without rind) of two cheese samples ($n = 6$).

Ninety non-trained members of staff and students performed sensory analysis, at the 30th, 60th and 90th days of ripening. Each sensory evaluation test involved 30 members which, individually, expressed their consent to participate in the tests. The tests involved evaluation of the cheese samples according to the following parameters: external and sliced aspect, aroma, taste and texture. Overall impression was also evaluated through a ranking test. Each cheese category was coded and the tasters were asked to evaluate both the visual and gustatory aspects using a 1–9 scale (1 = dislike extremely; 2 = dislike very much; 3 = dislike moderately; 4 = dislike slightly; 5 = neither like nor dislike; 6 = like slightly; 7 = like moderately; 8 = like very much; 9 = like extremely).

One-way ANOVA tests, included in StatSoft Statistica 8.0 (Statsoft Iberica, Lisbon, Portugal), were performed to compare the means of the physicochemical properties of the cheeses and the attributes used for sensory evaluation. The Tukey HSD post-hoc test, with a 95% confidence level was applied to assess differences between treatments.

3. Results and Discussion

Table 1 presents the composition of the different ingredients added as fat replacers to the 1.5% (*v/v*) fat milk batches. Significant differences were observed in the protein, fat and ash contents of those products. Concentrated buttermilk presented the lowest level of protein, while CS presented the highest value. The lowest fat content was observed in CS. CB and CW presented similar dry matter values, while CS showed a significantly lower content.

Table 1. Proximal composition of the different ingredients used for cheese production (% *w/v*).

Ingredients	Dry Matter	Protein	Fat	Ash
CB	13.78 [a] ± 0.42	3.65 [a] ± 0.07	1.41 [a] ± 0.01	0.75 [a] ± 0.09
CW	13.19 [a] ± 0.68	5.49 [b] ± 0.04	2.81 [b] ± 0.01	0.92 [b] ± 0.03
CS	10.35 [b] ± 0.14	6.56 [c] ± 0.05	0.41 [c] ± 0.01	0.54 [c] ± 0.02

(CB): UF concentrated buttermilk; (CW): UF concentrated whey; (CS): UF concentrated sheep's second cheese whey. Means within the same column with different superscripts are significantly different ($p < 0.05$).

The differences in the composition of the ingredients are reflected in the dry matter and fat content of the mixtures used for cheese production (Table 2). The mixtures containing CW and CS presented the lowest levels of fat, while the mixture used for the production of FF cheese presented a higher level of fat and lower levels of protein, lactose and minerals. The protein, lactose and ash contents did not show significant differences between the mixtures used for the production of RF cheeses.

Table 2. Composition of the different milk batches used for cheese production (% *w/v*).

Milk Batches	DM	F	P	L [1]	A
FFM	11.45 [a] ± 0.04	3.35 [a] ± 0.01	3.04 [a] ± 0.02	4.40 [a] ± 0.03	0.66 [a] ± 0.00
RFM	9.91 [b] ± 0.08	1.41 [b] ± 0.03	3.18 [b] ± 0.02	4.63 [b] ± 0.03	0.69 [b] ± 0.00
RFM + CB	10.03 [b] ± 0.04	1.43 [b] ± 0.02	3.22 [b] ± 0.01	4.68 [b] ± 0.02	0.70 [b] ± 0.00
RFM + CW	9.63 [c] ± 0.16	1.17 [c] ± 0.03	3.17 [b] ± 0.05	4.61 [b] ± 0.07	0.69 [b] ± 0.01
RFM + CS	9.84 [bc] ± 0.14	1.22 [bc] ± 0.05	3.23 [b] ± 0.03	4.69 [b] ± 0.05	0.70 [b] ± 0.01

(FFM): full-fat milk; (RFM): reduced-fat milk; (RFM + CB): reduced fat milk plus UF concentrated buttermilk; (RFM + CW): reduced fat milk pus UF concentrated whey; (RFM + CS): reduced fat milk plus UF concentrated sheep's second cheese whey; (DM): dry matter; (F): fat; (P): protein; (L): lactose; (A): ash. Means within the same column with different superscripts are significantly different ($p < 0.05$) ([1] calculated by difference).

As it can be observed in Figure 1a, in all stages of ripening, the FF cheese presented a significantly higher ($p < 0.05$) level of solids when compared to all other samples. Both the type of cheese and ripening time had significant effects on the dry matter content. Concerning moisture in non-fat solids (MNFS) FF cheeses also presented higher values (Figure 1b). Exceptions were the values of MNFS of RF, RF + CB and RF + CW at the 30th day of ripening. At the 60th day of ripening, RF + CB also presented values similar to FF.

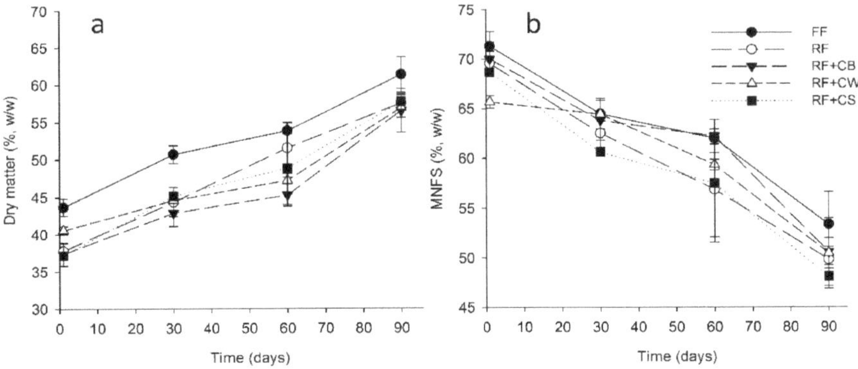

Figure 1. Dry matter (**a**) and moisture in non-fat solids (MNFS) (**b**) of tested cheeses over ripening. (FF): full-fat; (RF): reduced-fat; (RF + CB): reduced-fat with concentrated buttermilk; (RF + CW): reduced-fat with concentrated whey; (RF + CS): reduced-fat with concentrated second cheese whey.

According to the Portuguese standard, NP-1598 [31], at the 30th day of ripening all cheeses could be considered as semi-soft (61–69% MNFS). At the 60th day, only FF and RF + CB cheeses maintained this classification, being all others classified as semi-hard (54–63% MNFS). At the end of the ripening period (90th days), all the cheeses were classified as hard (49–56% MNFS).

The protein content was significantly lower in the FF cheeses (Figure 2a). In all other cheese samples, protein represented more than 50% of the solids, having the highest values been observed in RF + CW and RF + CS. Concerning fat content (Figure 2b), the FF cheese presented more than 45% fat on dry basis (being classified as a full-fat cheese according to NP-1598), whereas all other cheeses presented values ranging from 20 to 30%. In the cases of RF and of RF + CB the cheeses could be classified as half-fat (25–45% dry basis) in all stages of ripening. Cheeses with values of fat in the range 10–25% are classified as low-fat. This is the case of RF + CS in all stages of ripening and of RF + CW until the 60th day of ripening.

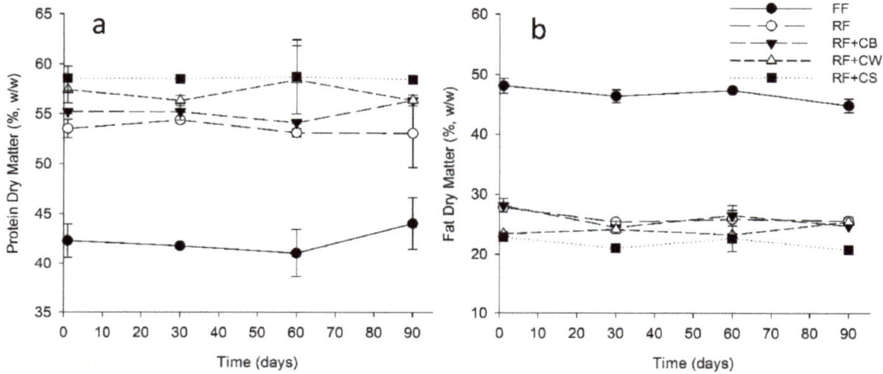

Figure 2. Protein in dry matter (**a**) and fat in dry matter (FDM) (**b**) of tested cheeses over ripening. (FF): full-fat; (RF): reduced-fat; (RF + CB): reduced-fat with concentrated buttermilk; (RF + CW): reduced-fat with concentrated whey; (RF + CS): reduced-fat with concentrated second cheese whey.

The ratio protein in dry matter/fat in dry matter (Pdm/Fdm) (Figure 3a) showed clear differences between FF and RF cheese samples. This value was lower than 1 in the FF cheese, whereas in the case of the remaining samples it was in the range of 1.8–2.8, this being the highest value observed in RF + CS. Although CS presented significantly higher protein content when compared to CB or CW (Table 1), the mixtures used for cheese production did not show significant differences regarding this parameter. Hence, the protein content of CS cannot justify, on its own, the higher Pdm/Fdm ratio observed in RF + CS. Second cheese whey normally presents a high proportion of whey protein aggregates resulting from the drastic heat treatment (*ca.* 90 °C 10 min) to which whey is submitted during the production of whey cheeses. The better retention of such aggregates in the cheese curd, as compared to native proteins, may explain the higher protein content of RF + CS and, to some extent, of RF + CW (in which protein was also denatured). With regard to the ratio protein in dry matter/moisture (Pdm/M), the maximum value attained by the FF cheeses was around 1.1, at the end of the ripening period, whereas in the case of the reduced fat cheeses was in the range of 1.2–1.4 at the end of ripening (Figure 3b). Higher values of Pdm/Fdm and of Pdm/M are expected to promote a firmer texture, often associated to the lower sensory scores obtained by RF cheeses. All the RF cheeses presented ratios of Pdm/M higher than 1.0 after the 30th of ripening with the exception of RF + CB cheeses that maintained Pdm/M values lower than 1.0 until the 60th day of ripening. This fact had positive repercussions on the textural and sensory properties of these cheeses, comparing well with FF cheese.

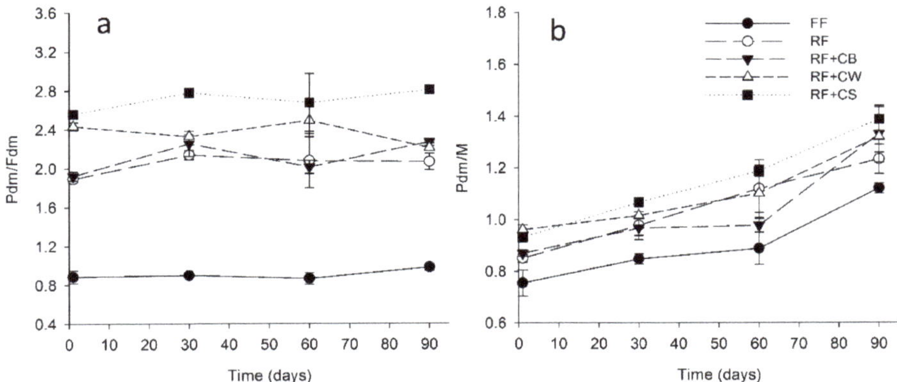

Figure 3. Ratio protein in dry matter/fat in dry matter (Pdm/Fdm) (**a**) and protein in dry matter/moisture (Pdm/M) (**b**) of tested cheeses over ripening. (FF): full-fat; (RF): reduced-fat; (RF + CB): reduced-fat with concentrated buttermilk; (RF + CW): reduced-fat with concentrated whey; (RF + CS): reduced-fat with concentrated second cheese whey.

On the first day of ripening, the pH values of the cheeses were in the order of 5.4–5.9, being significantly lower ($p < 0.05$) in the case of RF + CW and significantly higher in the case of RF + CS (ca. 5.9) (Figure 4a). After the 30th day of ripening, the values decreased to 5.0 in the case of FF and RF, being significantly higher in the cases of cheeses with added fat replacers (ca. 5.2). After this moment, the pH increased steadily until the end of ripening, the increase was significantly ($p < 0.05$) more pronounced in the case of cheeses with added fat replacers. The RF + CS cheeses showed higher pH values at the 60th day of ripening, the results were significantly higher on the 90th day. The titratable acidity (TA) showed an opposite tendency (Figure 4b). In this case, the highest values were observed for FF and RF cheeses at the 60th and 90th days of ripening. Overall, the reduced-fat cheeses containing fat replacers presented lower TA values, these being the lowest values observed for RF + CW in all stages of ripening.

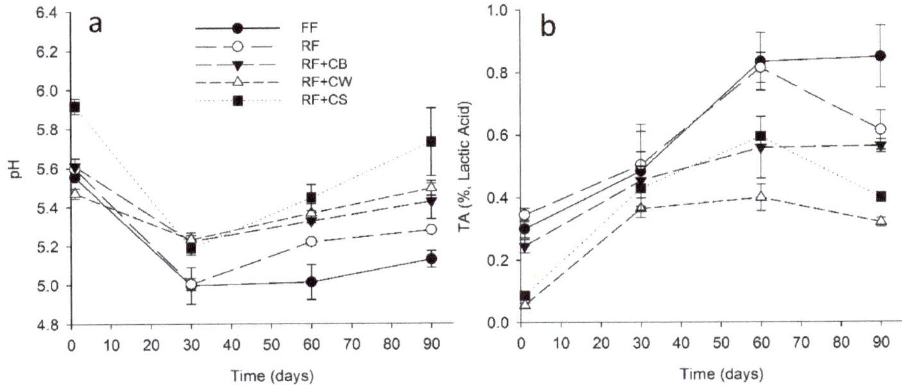

Figure 4. (**a**) pH and (**b**) titratable acidity (TA) of tested cheeses over ripening. (FF): full-fat; (RF): reduced-fat; (RF + CB): reduced-fat with concentrated buttermilk; (RF + CW): reduced-fat with concentrated whey; (RF + CS): reduced-fat with concentrated second cheese whey.

Concerning the colour parameters of the cheeses (Figure 5), the luminosity (L*) of the rind of RF + CW cheeses was significantly lower at the first day of ripening. At the 30th day, the L* value of the rind was significantly higher in RF and RF + CB when compared to all other samples. After the 60th day of ripening all the reduced-fat cheeses showed significantly lower L* values when compared to FF, having the cheeses with added fat replacers a clearly darker tone when compared to FF an RF. The luminosity of the paste showed a tendency to increase between the 30th and the 60th day, except in the case of RF + CS. At the end of the ripening period all cheeses showed significantly lower L* values when compared to the initial values, as a result of the progressive darkening of the paste. At the 90th day, all the reduced-fat cheeses showed significantly lower L* values of the paste, being the paste of RF + CW and RF + CS significantly darker than the ones of RF and RF + CB. Thus, it appears that both those fat replacers significantly impaired the colour of the cheeses paste. Concerning the a* parameter of the rind, the initial values were very similar, in short the initial white colour of the products shifted towards green (*ca.* −3.5) after 30 days. Then, these values increased until the 60th day, having the increase been more pronounced in RF and RF + CB. By the end of ripening the a* values of the rind decreased again in all reduced-fat cheeses, with the exception of RF + CW. At the end of ripening RF + CS showed a significantly lower a* value. The a* values of the paste were very similar until the 30th day, then decreased at the 60th day and finally increased slightly at the end of the ripening period. This increase was more pronounced in RF + CW and RF + CS. The FF cheese presented significantly lower a* values at the 90th day of ripening. From the 1st to the 30th days of ripening, the b* value of the rind shifted from 0 to around 20 in the case of the FF cheese, while in the cheeses with fat replacers, at the 30th day, the b* values were in the order of 14–16. This evolution indicates the shift from white to yellow. Then, the b* values were maintained, or slightly reduced, until the end of ripening. The same pattern was observed with the b* paste's value. However, the change was only significant from the 60th day onwards. The paste of the FF cheese presented a more pronounced ($p < 0.05$) yellow colour at the 60th and 90th days of ripening, indicating that fat has a significant impact on this parameter. RF cheeses presented significantly lower values of b* after the 60th day of ripening. RF + CW and RF + CS presented the lowest values, while RF and RF + CB presented intermediate values.

In regards to the texture (Figure 6), hardness values showed significant differences ($p < 0.05$) at the 30th day of ripening, having the RF + CW and RF + CS presented slightly higher values. From the 60th to the 90th day of ripening, a significant increase of the hardness values was observed in the cases of RF, RF + CW and RF + CS, being less pronounced in RF. FF cheeses and RF + CB maintained values of hardness in the order of 5.0–7.5 N during the entire ripening. In the case of adhesiveness, RF + CW and RF + CS also presented significantly lower values by the end of the ripening period, whereas RF + CB presented the highest values, although not significantly different from FF and RF. In all cases, this parameter significantly decreased between the 60th and the 90th day of ripening. At the end of ripening, chewiness values were significantly higher in RF, RF + CW and RF + CS being the values of RF + CB similar to those of FF. It is evident that, after the 60th day of ripening, with the exception of RF + CB, all the reduced-fat cheeses presented clear differences in these texture parameters when compared to FF cheeses. Thus, it can be considered that the use of liquid buttermilk was the best option for the replacement of fat, since RF + CB cheeses are very similar to FF cheeses. Other authors reported that fat reduction increased the hardness of *Minas* fresh cheeses, promoting a denser microstructure and less proteolysis [14]. In a previous work we reported values in the order of 3.6 N for the hardness of RF cheeses with addition of 10 (*v/v*) liquid whey protein concentrates (LWPC) plus 0.25–0.5% (*m/v*) Simplesse™, while the conventional RF cheeses showed values in the order of 8 N. The cheeses with addition of LWPC also presented significantly higher levels of MNFS, when compared to conventional RF cheeses [25]. As reported by other authors, cheese fracturability and hardness increase with decreasing fat, while elasticity and adhesiveness decrease. Cheese lightness and red and yellow indexes also decrease with decreasing fat content, as it occurred with our samples [32].

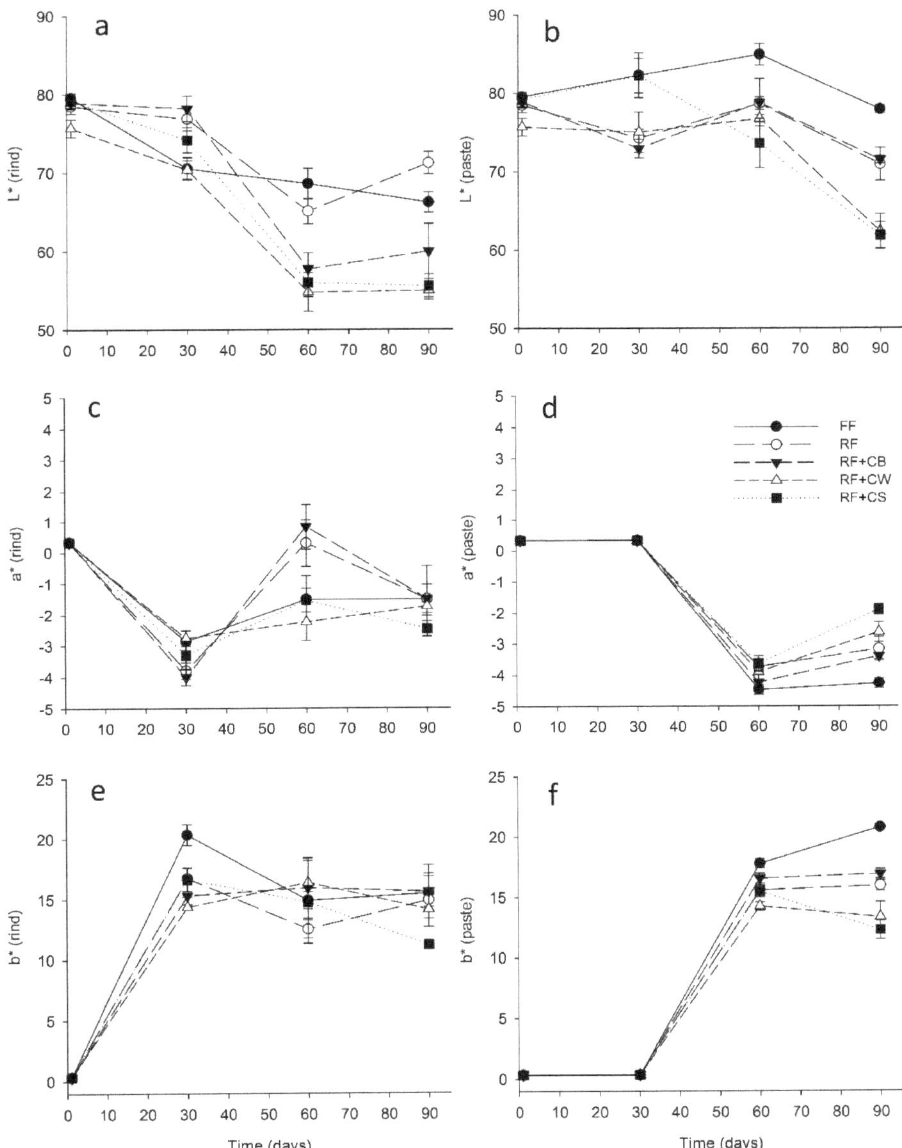

Figure 5. Colour parameters of the rind (**a**,**c**,**e**) and of the paste (**b**,**d**,**f**) of tested cheeses over ripening. (FF): full-fat; (RF): reduced-fat; (RF + CB): reduced-fat with concentrated buttermilk; (RF + CW): reduced-fat with concentrated whey; (RF + CS): reduced-fat with concentrated second cheese whey.

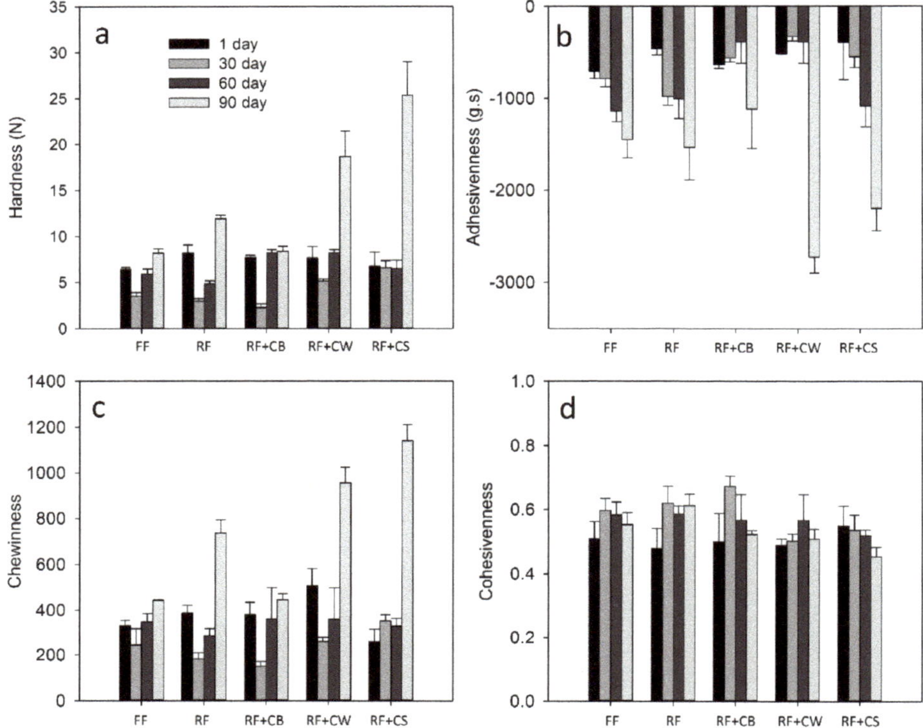

Figure 6. Texture parameters of the tested cheeses over ripening. (**a**) Hardness; (**b**) Adhesivenness; (**c**) Chewinness; (**d**) Cohesivenness. (FF): full-fat; (RF): reduced-fat; (RF + CB): reduced-fat with concentrated buttermilk; (RF + CW): reduced-fat with concentrated whey; (RF + CS): reduced-fat with concentrated second cheese whey.

The sensory evaluation results are depicted in Tables 3 and 4. At the 30th day of ripening no differences between cheeses could be detected with regards to their appearance. RF + CW and RF + CS cheeses obtained significantly lower scores for texture and taste. This fact is also demonstrated by the lower ranking obtained by both samples (Table 4). At the 60th day of ripening the defects in texture and taste of RF + CW were not evident, while RF + CS showed significantly lower scores for these parameters. However, at the end of the ripening period RF + CW presented significantly lower scores for appearance, texture and taste. The FF cheese presented the highest scores for aroma and taste, at the 60th day of ripening, and for appearance and taste, at the end of the ripening period. RF + CB presents similar results to the FF cheeses in all stages of ripening. It should be highlighted that, at the 30th day of ripening, RF + CB cheeses obtained the highest scores, although not significantly different from the ones obtained by FF cheeses. The main defect reported by panelists was related to the hardness of RF + CW and RF + CS at the 90th day.

Table 3. Sensory evaluation of the cheese samples at the 30th, 60th and 90th days of ripening.

Cheese	AP 30	AR 30	TE 30	TA 30
FF	7.00 [a] ± 1.17	6.63 [ab] ± 1.85	7.03 [a] ± 1.22	7.23 [a] ± 1.63
RF	7.10 [a] ± 1.71	6.63 [ab] ± 1.56	6.67 [ab] ± 1.73	6.57 [ab] ± 1.83
RF + CB	7.50 [a] ± 1.17	6.87 [a] ± 1.38	7.33 [a] ± 1.49	7.27 [a] ± 1.26
RF + CW	6.80 [a] ± 1.65	6.00 [ab] ± 1.60	5.77 [b] ± 1.45	5.73 [b] ± 1.98
RF + CS	6.63 [a] ± 1.50	5.73 [b] ± 1.48	5.57 [b] ± 1.65	5.57 [b] ± 1.94
	AP 60	**AR 60**	**TE 60**	**TA 60**
FF	7.53 [a] ± 1.20	7.23 [a] ± 1.10	7.53 [ab] ± 1.20	7.50 [a] ± 1.28
RF	7.30 [a] ± 0.95	6.70 [a] ± 1.42	7.40 [ab] ± 1.35	7.27 [ab] ± 1.55
RF + CB	7.53 [a] ± 0.94	7.03 [a] ± 1.19	7.67 [a] ± 1.18	7.10 [ab] ± 1.27
RF + CW	7.23 [a] ± 1.04	6.77 [a] ± 1.38	7.00 [ab] ± 1.39	6.90 [ab] ± 1.32
RF + CS	7.20 [a] ± 1.19	6.60 [a] ± 1.48	6.63 [b] ± 1.73	6.43 [b] ± 1.89
	AP 90	**AR 90**	**TE 90**	**TA 90**
FF	7.77 [a] ± 1.28	7.13 [a] ± 1.61	7.37 [a] ± 1.69	7.50 [a] ± 1.59
RF	7.40 [ab] ± 1.04	6.70 [a] ± 1.58	6.90 [ab] ± 1.54	6.93 [ab] ± 1.53
RF + CB	7.43 [ab] ± 0.94	6.90 [a] ± 1.40	6.97 [ab] ± 1.63	7.00 [ab] ± 1.51
RF + CW	5.93 [b] ± 1.66	6.17 [a] ± 2.02	6.00 [b] ± 1.82	6.33 [b] ± 1.63
RF + CS	6.77 [ab] ± 1.36	6.10 [a] ± 1.90	6.63 [ab] ± 1.52	6.63 [ab] ± 1.52

(AP) = appearance; (AR) = aroma; (TE) = texture; (TA) = taste. (FF) full-fat; (RF) reduced-fat; RF + CB: reduced-fat with concentrated buttermilk; RF + CW: reduced-fat with concentrated whey; RF + CS: reduced-fat with concentrated second cheese whey. Means within the same column with different superscripts are significantly different ($p < 0.05$).

Table 4. Ranking of the cheese samples at the different periods of ripening. Lower values indicate higher positioning in the ranking.

Cheese	Rank 30	Rank 60	Rank 90
FF	2.23 [a] ± 1.30	2.23 [a] ±1.22	2.17 [a] ± 1.49
RF	2.80 [a] ± 1.32	2.97 [ab] ± 1.30	2.60 [ab] ± 1.35
RF + CB	2.23 [a] ±1.07	2.60 [ab] ± 1.28	2.90 [ab] ± 1.18
RF + CW	3.83 [b] ± 1.23	3.50 [b] ± 1.46	3.87 [b] ± 1.31
RF + CS	3.90 [b] ± 1.18	3.70 [b] ± 1.37	3.47 [b] ± 1.14

(FF): full-fat; (RF): reduced-fat; (RF + CB): reduced-fat with concentrated buttermilk; (RF + CW): reduced-fat with concentrated whey; (RF + CS): reduced-fat with concentrated second cheese whey. Means within the same column with different superscripts are significantly different ($p < 0.05$). (AP) = appearance; (AR) = aroma; (TE) = texture; (TA) = taste. Means within same column with different superscripts are significantly different ($p < 0.05$).

At the 60th day of ripening the ratio Pdm/M of RF + CB showed values similar to those of FF cheeses, while in all other cases this ratio presented significantly higher values, which adversely affected texture. It is reported by Skeie et al., 2013 that microparticulated whey protein and buttermilk added to the cheese milk improved the texture of RF cheeses, whereas the flavour could be improved by selected *Lactobacillus* spp. isolated from good-quality cheeses [33]. The results obtained by these authors showed that it was possible to produce a 10% fat Dutch cheese with an improved texture compared with the regular cheese without any additional ingredients. MWP also improved yield and the textural properties of RF Cheddar cheese due to the water-binding ability of denatured whey protein and by decreasing firmness [34]. Hence, similar results could be expected with the addition of concentrated cheese whey (CW). However, this was not observed, particularly after the 60th day of ripening. Perreault and co-workers assessed the effect of denatured whey protein concentrate (DWPC) and its fractions on cheese yield composition, and rheological properties of cheeses. For cheeses with the same moisture content, the use of DWPC had no direct effect on rheological parameters. The protein aggregates were primarily responsible for the increase in cheese yield while moisture content explained, to a large extent, the variation in cheese rheological properties [35]. Other authors evaluated the fat mimicking mechanism of MWP in milk-based systems using rheological and tribological techniques, and reported that friction levels attained with MPW proteins and dairy fat at typical speeds involved in oral processing were comparable, demonstrating therefore the capability of MWP dispersions to imitate dairy fat in milk-based systems from a lubrication point of view [36].

Regarding the use of buttermilk powder in LF Cheddar cheese, it is reported that cheese made with BM addition had a homogeneous protein network with small voids and a smoother and less coarse structure when compared to LF cheeses without buttermilk addition [37]. The addition of liquid BM to cheese milk was also tested. As the percentage of BM increased, the total solids, fat, protein, fat in dry matter and ash of cheese milk decreased significantly, leading to a softer and moister curd. However, samples prepared with more than 25% BM were not acceptable with respect to the taste panel [38]. The effects of BM powder addition post-curd formation, or liquid BM addition to cheese milk on the characteristics of Cheddar-style cheese were evaluated in parallel. Addition of 10% BM powder resulted in higher phospholipid content, moisture, pH and salt levels, and lower fat in dry matter. BM addition also originated a more porous cheese microstructure with higher fat globule coalescence and increased free fat, while increasing moisture and decreasing protein, fat and pH levels [39]. It is also reported that liquid BM addition to cheese milk resulted in a softer cheese compared to other cheeses, while BM powder addition had no influence on cheese firmness compared to the control cheese. However, significant differences in sensory profiles associated with off-flavour were also observed with the addition of liquid BM to cheese milk. Addition of 10% BM powder to cheese curds resulted in cheese comparable to the control Cheddar with similar structural and sensory characteristics, although with differences in overall cheese flavour [40]. In the case of our products no adverse effects resulted from the addition of BM to cheese milk and this fact may be attributed to the lower amount added.

4. Conclusions

The reduction of fat in cheeses often affects negatively their sensory properties. Therefore, several approaches are normally used to minimize those negative effects. In the present study, UF concentrated liquid buttermilk, whey protein concentrate, and sheep's second cheese whey were used for this purpose. From the results obtained, it is evident that UF concentrated liquid buttermilk significantly improved the properties of RF/LF cheeses, which showed good overall sensory evaluation and compared well to FF cheeses. The use of UF concentrated dairy by-products can allow for their direct valorisation in dairy plants and represents a significant contribution to the circular economy. It is recommended that further work should compare the fat replacing properties of such products, both in the liquid and dry form. Optimization of mixtures of such by-products should also deserve further research.

Author Contributions: Investigation, formal analysis and methodology A.R.B., A.F.P., N.G.M., and D.G.G.; writing—review and editing, supervision M.F.H.; supervision, writing—original draft preparation, funding acquisition and project administration C.D.P. All authors have read and agreed to the published version of the manuscript.

Funding: This research was funded by national funds through the Ministry of Agriculture and Rural Development and co-financed by the European Agricultural Fund for Rural Development (EAFRD), through the partnership agreement Portugal 2020-PDR, under the project PDR2020-101-030768: LACTIES-*Inovação, Ecoeficiência e Segurança em PME´s do Sector dos Lacticínios;* and through the Portuguese Foundation for Science and Technology (FCT): project UID/AMB/00681/2019.

Acknowledgments: The authors are grateful to Adélia Vaz, Lurdes Pires and Jorge Arede for their help in the production of cheeses in pilot plant.

Conflicts of Interest: The authors declare no conflict of interest.

References

1. Rogers, N.R.; Drake, M.A.; Daubert, C.R.; Mcmahon, D.J.; Bletsch, T.K.; Foegeding, E.A. The effect of aging on low-fat, reduced-fat, and full-fat Cheddar cheese texture. *J. Dairy Sci.* **2009**, *92*, 4756–4772. [CrossRef] [PubMed]
2. Zalazar, C.A.; Zalazar, C.S.; Bernal, S.; Bertola, N.; Bevilacqua, A.; Zraitzky, N. Effect of moisture level and fat replacer on physico-chemical, rheological and sensory properties of low fat soft cheeses. *Int. Dairy J.* **2002**, *12*, 45–50. [CrossRef]
3. McCarthy, C.M.; Wilkinson, M.G.; Kelly, P.M.; Guinee, T.P. Effect of salt and fat reduction on proteolysis, rheology and cooking properties of Cheddar cheese. *Int. Dairy J.* **2016**, *56*, 74–86. [CrossRef]
4. Banks, J.M. The technology of low-fat cheese manufacture. *Int. J. Dairy Technol.* **2004**, *57*, 199–208. [CrossRef]
5. Drake, M.A.; Miracle, R.E.; Mcmahon, D.J. Impact of fat reduction on flavour and flavour chemistry of Cheddar cheeses. *J. Dairy Sci.* **2010**, *93*, 5069–5081. [CrossRef]
6. Rogers, N.R.; McMahon, D.J.; Daubert, C.R.; Berry, T.K.; Foegeding, E.A. Rheological properties and microstructure of Cheddar cheese made with different fat contents. *J. Dairy Sci.* **2010**, *93*, 4565–4576. [CrossRef]
7. Mistry, V.V. Low fat cheese technology. *Int. Dairy J.* **2001**, *11*, 413–422. [CrossRef]
8. Mayta-Hancco, J.; Trujillo, A.J.; Zamora, A.; Juan, B. Effect of ultra-high pressure homogenisation of cream on the physicochemical and sensorial characteristics of fat-reduced starter-free fresh cheeses. *LWT Food Sci. Technol.* **2019**, *110*, 292–298. [CrossRef]
9. Di Cagno, R.; De Pasquale, I.; De Angelis, M.; Buchin, S.; Rizzello, C.G.; Gobbetti, M. Use of microparticulate whey protein concentrate, exopolysaccharide-producing Streptococcus thermophilus, and adjunct cultures for making low-fat Italian Caciotta-type cheese. *J. Dairy Sci.* **2014**, *97*, 72–84. [CrossRef] [PubMed]
10. Oluk, A.C.; Güven, M.; Hayaloglu, A.A. Proteolysis texture and microstructure of low-fat Tulum cheese affected by exopolysaccharide-producing cultures during ripening. *Int. J. Food Sci. Technol.* **2014**, *49*, 435–443. [CrossRef]
11. Oluk, A.C.; Güven, M.; Hayaloglu, A.A. Influence of exopolysaccharide-producing cultures on the volatile profile and sensory quality of low-fat Tulum cheese. *Int. J. Dairy Technol.* **2014**, *67*, 265–276. [CrossRef]
12. Koca, N.; Metin, M. Textural, melting and sensory properties of low-fat fresh Kashar cheeses produced by using fat replacers. *Int. Dairy J.* **2004**, *14*, 365–373. [CrossRef]
13. Lashkari, H.; Khosrowshahi, A. Chemical composition and rheology of low-fat Iranian white cheese incorporated with guar gum and gum arabic as fat replacers. *J. Food Sci. Technol.* **2014**, *51*, 2584–2591. [CrossRef]
14. Diamantino, V.R.; Beraldo, F.A.; Sunakozawa, T.N.; Lúcia, A.; Penna, B. Effect of octenyl succinylated waxy starch as a fat mimetic on texture, microstructure and physicochemical properties of Minas fresh cheese. *LWT Food Sci. Technol.* **2014**, *56*, 356–362. [CrossRef]
15. Wydrych, J.; Gawron, A.; Jeli, T. The effect of fat replacement by inulin on the physicochemical properties and microstructure of acid casein processed cheese analogues with added whey protein polymers. *Food Hydrocoll.* **2015**, *44*, 1–11. [CrossRef]
16. Wang, F.; Tong, Q.; Luo, J.; Xu, Y.; Ren, F. Effect of carrageenan on physicochemical and functional properties of low-fat Colby cheese. *J. Food Sci.* **2016**, E1949–E1955. [CrossRef]
17. Palatnik, D.R.; Herrera, P.A.; Rinaldoni, A.N.; Basurto, R.O.; Campderrós, M.E. Development of reduced-fat cheeses with the addition of Agave fructans. *Int. J. Dairy Technol.* **2016**, *70*, 212–219. [CrossRef]
18. Dai, S.; Jiang, F.; Corke, H.; Shah, N.P. Physicochemical and textural properties of mozzarella cheese made with konjac glucomannan as a fat replacer. *Food Res. Int.* **2018**, *107*, 691–699. [CrossRef]
19. Sharma-Khanal, B.K.; Bhandari, B.; Prakash, S.; Liu, D.; Zhou, P.; Bansal, N. Modifying textural and microstructural properties of low fat Cheddar cheese using sodium alginate. *Food Hydrocoll.* **2018**, *83*, 97–108. [CrossRef]
20. Li, H.; Liu, Y.; Sun, Y.; Li, H.; Yu, J. Properties of polysaccharides and glutamine transaminase used in mozzarella cheese as texturizer and crosslinking agents. *LWT Food Sci. Technol.* **2019**, *99*, 411–416. [CrossRef]
21. Torres, I.C.; Mutaf, G.; Larsen, F.H.; Ipsen, R. Effect of hydration of microparticulated whey protein ingredients on their gelling behaviour in a non-fat milk system. *J. Food Eng.* **2016**, *184*, 31–37. [CrossRef]

22. Zhang, Z.; Arrighi, V.; Campbell, L.; Lonchamp, J. Properties of partially denatured whey protein products: Formation and characterisation of structure. *Food Hydrocoll.* **2016**, *52*, 95–105. [CrossRef]

23. Cheftel, J.C.; Dumay, E. Microcoagulation of proteins for development of creaminess. *Food Rev. Int.* **1993**, *9*, 473–502. [CrossRef]

24. Liu, K.; Stieger, M.; Van der Linden, E.; Van de Velde, F. Effect of microparticulated whey protein on sensory properties of liquid and semi-solid model foods. *Food Hydrocoll.* **2016**, *60*, 186–198. [CrossRef]

25. Henriques, M.; Gomes, D.; Brennan, K.; Skryplonek, K.; Fonseca, C.; Pereira, C. The use of whey proteins as fat replacers for the production of reduced fat cheeses. In *Cheese Production, Consumption and Health Benefits*; Henriques, M., Pereira, C., Eds.; Nova Science Publishers: New York, NY, USA, 2018; pp. 139–168.

26. Hinrichs, J. Incorporation of whey proteins in cheese. *Int. Dairy J.* **2001**, *11*, 495–503. [CrossRef]

27. Frusch, J.A.H.; Kokx, J.J.M.P. Cheese with Added Whey Protein Agglomerates. European Patent Application EP1917861 A1, 7 May 2008.

28. Giroux, H.J.; Veilllete, N.; Britten, M. Use of denatured whey protein in the production of artisanal cheeses from cow, goat and sheep milk. *Small Rum. Res.* **2018**, *161*, 34–42. [CrossRef]

29. AOAC. *AOAC Official Methods of Analysis of Association of Official Analytical Chemists*, 16th ed.; 33 Dairy Products; AOAC: Rockville, MD, USA, 1997; Volume II.

30. ISO. *ISO 3433 Cheese-Determination of Fat Content—Van Gulik Method*; International Organization for Standardization: Geneva, Switzerland, 2008.

31. IPQ. *IPQ-Portuguese Institute of Quality (NP-1598)—Cheese: Definition, Classification, Packaging and Marking*; IPQ: Lisbon, Portugal, 1983.

32. Kavas, G.; Oysun, G.; Kinik, O.; Uysal, H. Effect of some fat replacers on chemical, physical and sensory attributes of low-fat white pickled cheese. *Food Chem.* **2004**, *88*, 381–388. [CrossRef]

33. Skeie, S.; Alseth, G.M.; Østlie, H.; Abrahamsen, R.K.; Johansen, A.G.; Øyaas, J. Improvement of the quality of low-fat cheese using a two-step strategy. *Int. Dairy J.* **2013**, *33*, 153–162. [CrossRef]

34. Stankey, J.A.; Lu, Y.; Abdalla, A.; Govindasamy-Lucey, S.; Jaeggi, J.J.; Mikkelsen, B.Ø.; Kenneth, T.; Pedersen, K.T.; Andersen, C.B. Low-fat Cheddar cheese made using microparticulated whey proteins: Effect on yield and cheese quality. *Int. J. Dairy Technol.* **2017**, *70*, 481–491. [CrossRef]

35. Perreault, V.; Rémillard, N.; Chabot, D.; Morin, P.; Pouliot, Y.; Britten, M. Effect of denatured whey protein concentrate and its fractions on cheese composition and rheological properties. *J. Dairy Sci.* **2017**, *100*, 5139–5152. [CrossRef]

36. Olivares, M.L.; Shahrivar, K.; de Vicente, J. Soft lubrication characteristics of microparticulated whey proteins used as fat replacers in dairy systems. *J. Food Eng.* **2019**, *245*, 157–165. [CrossRef]

37. Romeih, E.A.; Moe, K.M.; Skeie, S. The influence of fat globule membrane material on the microstructure of low-fat Cheddar cheese. *Int. Dairy J.* **2012**, *26*, 66–72. [CrossRef]

38. Bahrami, M.; Ahmadi, D.; Beigmohammadi, F.; Hosseini, F. Mixing sweet cream buttermilk with whole milk to produce cream cheese. *Irish J. Agric. Food Res.* **2015**, *54*, 73–78. [CrossRef]

39. Hickey, C.D.; Diehl, B.W.K.; Nuzzo, M.; Millqvist-Feurby, A.; Wilkinson, M.G.; Sheehan, J.J. Influence of buttermilk powder or buttermilk addition on phospholipid content, chemical and bio-chemical composition and bacterial viability in Cheddar style cheese. *Food Res. Int.* **2017**, *102*, 748–758. [CrossRef]

40. Hickey, C.D.; O'Sullivan, M.G.; Davis, J.; Scholz, D.; Kilcawley, K.N.; Wilkinson, M.G.; Sheehan, J.J. The effect of buttermilk or buttermilk powder addition on functionality, textural, sensory and volatile characteristics of Cheddar-style cheese. *Food Res. Int.* **2018**, *103*, 468–477. [CrossRef]

© 2020 by the authors. Licensee MDPI, Basel, Switzerland. This article is an open access article distributed under the terms and conditions of the Creative Commons Attribution (CC BY) license (http://creativecommons.org/licenses/by/4.0/).

Article

Application of the UHPLC-DIA-HRMS Method for Determination of Cheese Peptides

Georg Arju [1,2,*], Anastassia Taivosalo [2], Dmitri Pismennoi [1,2], Taivo Lints [1,2], Raivo Vilu [2], Zanda Daneberga [3], Svetlana Vorslova [3], Risto Renkonen [4,5] and Sakari Joenvaara [4,5]

[1] Department of Chemistry and Biotechnology, School of Science, Tallinn University of Technology, Ehitajate tee 5, 12616 Tallinn, Estonia; dmitri@tftak.eu (D.P.); taivo.lints@tftak.eu (T.L.)

[2] Center of Food and Fermentation Technologies, Akadeemia tee 15A, 12618 Tallinn, Estonia; anastassia@tftak.eu (A.T.); raivo@tftak.eu (R.V.)

[3] Institute of Oncology, Riga Stradins University, 13 Pilsonu Str., LV-1002 Riga, Latvia; zanda.daneberga@rsu.lv (Z.D.); svetlana.vorslova@rsu.lv (S.V.)

[4] Transplantation Laboratory, Haartman Institute, University of Helsinki, FI-00014 Helsinki, Finland; risto.renkonen@helsinki.fi (R.R.); sakari.joenvaara@helsinki.fi (S.J.)

[5] HUSLAB, Helsinki University Hospital, FI-00029 Helsinki, Finland

* Correspondence: georg@tftak.eu; Tel.: +372-53-401-565

Received: 30 May 2020; Accepted: 22 July 2020; Published: 23 July 2020

Abstract: Until now, cheese peptidomics approaches have been criticised for their lower throughput. Namely, analytical gradients that are most commonly used for mass spectrometric detection are usually over 60 or even 120 min. We developed a cheese peptide mapping method using nano ultra-high-performance chromatography data-independent acquisition high-resolution mass spectrometry (nanoUHPLC-DIA-HRMS) with a chromatographic gradient of 40 min. The 40 min gradient did not show any sign of compromise in milk protein coverage compared to 60 and 120 min methods, providing the next step towards achieving higher-throughput analysis. Top 150 most abundant peptides passing selection criteria across all samples were cross-referenced with work from other publications and a good correlation between the results was found. To achieve even faster sample turnaround enhanced DIA methods should be considered for future peptidomics applications.

Keywords: dairy product analysis; cheese peptidomics; cheesemaking; data-independent acquisition

1. Introduction

During cheese ripening, caseins undergo a progressive breakdown by enzymatic action, releasing peptides and amino acids, which contributes to the development of cheese flavour and texture [1]. The term "peptidomics" has been extensively used in dairy science for comprehensive analysis of peptides released during proteolysis in different cheese varieties [2–4] as well as characterisation of bioactive peptides with potential nutritional and health-promoting effects [5,6]. Several researchers have been focusing on the identification of phosphorylated peptides [7] and the determination of specific bitter peptides and their contribution to cheese flavour [8,9]. Many studies have been carried out to evaluate the effect of different adjunct cultures on the formation of peptides in cheese and thus to adjust the taste and aroma characteristics of a final product [10,11].

The key analytical tool employed in cheese peptidome research (i.e., increasing the knowledge of proteolytic events occurring during ripening) as well as exploring the possibilities of controlling the cheese maturation process, is currently mass spectrometry (MS) coupled with liquid chromatography (LC) [12–14]. The most widely used hyphenation is nano ultra (high) performance liquid chromatography (nanoUHPLC) coupled with high-resolution mass spectrometry (HRMS) based on data-dependent acquisition (DDA) mode [15–17].

The samples complexity combined with the slow acquisition rate of DDA modes as well as the ever-growing demand for higher protein coverage typically results in analytical gradients that exceed 60 or even 120 min, making such methods less appealing for higher-throughput studies [4,14]. Data-independent acquisition (DIA) is an alternative acquisition mode to DDA. DIA, unlike DDA, does not rely on precursor isolation. DIA is based on a principle of a rapid alternation between low and high collision energies to acquire MS^1 and MS^2 spectra. DIA relies on a chromatographic alignment of MS^1 and MS^2 for fragment-precursor assignment. Operating at higher acquisition rates and being compatible with faster gradients, DIA has been employed in several food research applications such as food safety, authenticity testing and peptide profiling of various food matrices [18,19]. Using DIA it is possible to simultaneously acquire both qualitative and quantitative data.

The aim of this study was to develop an LC-DIA-MS-based methodology for cheese peptide profiling with a sub-60 min analytical gradient without compromises in chromatographic performance and protein coverage.

2. Materials and Methods

2.1. Materials

Hi3 EColi STD (p/n: 186006012) and [Glu1]-Fibrinopeptide (p/n: 700004729) were purchased from Waters Corporation (Milford, MA, USA). Peptide quantitation was performed using Pierce™ Quantitative Colorimetric Peptide Assay (C/N: 23275, Thermo Fisher Scientific, Waltham, MA, USA). Nanosep® Centrifugal Devices with Omega™ membrane 3 K were obtained from Pall (p/n: OD003C34, Port Washington, NY USA). Ultrapure water (18.2 MΩ.cm) was prepared with MilliQ® Direct-Q® UV (Merck KGaA, Darmstadt, Germany). Acetonitrile (MeCN; LiChrosolv, hypergrade for LC-MS,) and formic acid (FA; LC-MS grade) were acquired from Sigma-Aldrich (Darmstadt, Germany).

2.2. Cheese Manufacture and Sampling

Three Gouda-type cheeses (Cheese 1, Cheese 2 and Cheese 3) were produced using three different DL-starter cultures (DL1 and DL2 by Chr. Hansen Ltd., Hørsholm, Denmark and DL3 by DuPont™ Danisco®, Copenhagen, Denmark) at a dairy plant from 600 L of pasteurised (at 74 °C for 15 s) milk. Animal rennet (25 mL 100 1/L; 230 IMCU 1/g; 20/80) of chymosin and bovine pepsin, (Chr. Hansen Ltd., Hørsholm, Denmark) was added to milk. After coagulation, the curd was cut, whey removed, and cheese grains stirred and heated at 32 °C for 30 min. Cheeses were prepressed under whey, moulded, pressed for 1.5 h, brine salted (pH 5.1) for 36 h, waxed and ripened at 10–15 °C for 90 days. Samples were taken from each cheese at 0 (after salting), 14, 30 and 90 days of ripening, grated and stored at −20 °C for further analysis.

2.3. Sample Preparation

To prepare water-soluble extracts (WSE) of cheeses, 2.5 g of grated cheese was homogenised in 22.5 mL of MilliQ® water (12,500–13,000 rpm) using Polytron PT 2100 dispersing aggregate with a diameter of 20 mm (Kinematica AG, Switzerland). Samples were heated for 10 min at 75 °C and centrifuged for 20 min at 4 °C at 13,302 g. Supernatants were stored in 1.5 mL Eppendorf® Protein LoBind microcentrifuge tubes (Eppendorf AG, Germany) at −20 °C until further purification. Thawed sample supernatants (250 μL) and MilliQ® water (250 μL) were transferred into the Eppendorf® tube and vortexed for 30 s. Obtained mixture (400 μL) was added to Nanosep® with 3 K Omega™ spin filter. Samples were centrifuged at 11,200 g for 15 min. For peptide quantification 30 μL of filtrate was mixed with 970 μL of working reagent. After 30 min incubation at room temperature, absorbance was measured at 480 nm (Helios Gamma, Thermo Electron Corporation, Waltham, MA, USA) and concentrations calculated using a blank-adjusted calibration curve. For HRMS analysis, 30 μL of filtrate was mixed with 970 μL of MilliQ® water (1% MeCN and 0.1% FA). Using results from the peptide concentration measurement samples were further diluted to result in 100 ng/μL equalising column load across all the samples. The sample (195 μL) was transferred into a vial and spiked with 5 μL of 1 pmol/μL of Hi3 EColi STD.

2.4. Liquid Chromatography Mass Spectrometry

Samples were analysed using Waters nanoAcquity UPLC® system (Waters Corporation, Milford, MA, USA) coupled with a Waters MALDI SYNAPT G2-Si Mass Spectrometer equipped with NanoLockSpray Exact Mass Ionisation Source and controlled by Waters MassLynx 4.1 (V4.1 SCN916, Waters Corporation, Milford, MA). Mobile phases were as follows: (A) MilliQ® + 0.1% formic acid and (B) MeCN + 0.1% formic acid. Injection volume was 2 μL. Samples were loaded onto Acquity UPLC® Symmetry C18 Nanoacquity 10 k 2 g V/M Trap column (100A, 5 μm, 180 μm × 20 mm, Waters Corporation, Milford, MA, USA). Loading was carried out for 1 min at 5 μL/min using 1% B. Loaded sample was further analysed using Acquity UPLC® M-Class HSS T3 Column (1.8 μm, 75 μm × 150 mm, Waters Corporation, Milford, MA, USA) kept at 40 °C. The gradient was as follows: 0–1 min hold at 1% B, 1–10 min linear gradient 1–15% B, 10–40 min linear gradient 15–35% B, 40–45 min linear gradient 35–95% B, 45–53 min hold at 95% B, 53–55 min linear gradient 95–1% B, 55–70 min hold at 1% B. Analytical flow rate was 0.3 μL/min.

The instrument was operated in positive polarity and resolution mode (35000 FWHM at 785.8426 *m/z*). Data were acquired in MS^E mode with a scan time of 0.5 s between 1 and 50 min. Recorded mass range was from 50 to 2000 *m/z* for both low and high energy spectra. The collision energy was ramped from 15 to 45 V in the trap cell of the instrument. Cone voltage was set to 40 V and capillary voltage was set to 2.4 kV. Source offset was 60, source temperature was 80 °C. Cone gas was 50 L/h, nano flow gas was 0.3 bar and purge gas was 100 L/h. [Glu1]-Fibrinopeptide was used as LockMass for mass axis correction and was acquired every 30 s.

2.5. Raw Data Processing

The raw data files were imported to the Progenesis QI for proteomics software (Nonlinear Dynamics, Newcastle, UK). During the import masses were lock mass corrected with 785.8426 *m/z*, corresponding to doubly charged [Glu1]-Fibrinopeptide B. Default parameters for peak picking and alignment algorithm were used.

The peptides were searched against beta-casein (β-CN; P02666), alpha $_{s1}$-casein (α_{s1}-CN; P02662) and alpha $_{s2}$-casein (α_{s2}-CN; P02663) sequences from bovine species obtained with the UniProt database [20].

The protein identifications were done against sequences added with a spike in Hi3 standard ClpB protein sequence, CLPB_ECOLI (P63285). Nonspecific cleavage was chosen and zero missed cleavages were allowed. Fragment and peptide error tolerances were set to auto and false discovery rate to <1%. One or more fragment ions per peptide were required for ion matching. The following variable post-translational modifications were used in the analysis: oxidation (M), acetyl-(protein N-terminal), deamidation (NQ) and phosphorylation (STY). The analysis of each post-translational modification was done separately, and the results were combined. Absolute mass error for a peptide was set to 5 ppm and we included peptides with one to three charges in the analysis. In the sample grouping, the within-subject design was used, fold changes and repeated measures of ANOVA were used for statistics. Filtered data were exported and then subjected to the normalisation of peptide abundances based on the coefficients of each sample dilution.

2.6. Data Analysis

An additional batch of samples was analysed using the methodology described by Taivosalo et al. [14] to highlight differences in results between two approaches. DDA experiment raw data files were imported to MaxQuant proteomics software (https://www.maxquant.org/) for data analysis as described in the publication and subsequently exported for intramethod correlation analysis. Filtered data were exported and then subjected to the normalisation of peptides abundances based on the coefficients of each sample dilution. Normalised abundances were used to construct a data matrix to identify differences between sample peptide compositions.

The comparison of the DIA and DDA methods was done with the help of in-house data analysis and visualisation scripts written in the Python™ programming language (Python Software Foundation). For both methods for each measured sample, the top 150 peptides with the highest intensities were found. The locations of those peptides were then found on the protein sequences the peptides originated from, and peptide coverage profiles were created for each casein in every sample for both methods, showing the peptides coloured by the intensity and laid out on their corresponding protein sequences.

3. Results

Overall, 558 peptides were identified among the analysed samples across 90 days of ripening using our method. The variation in peptide profiles and abundances was evaluated.

Ten per cent of samples (Day 0 of each cheese) were injected as triplicates. Relative standard deviation for all replicates equated to 10.88%.

It was found that at Day 90, Cheese 2 had the lowest number of peptides (Figure 1).

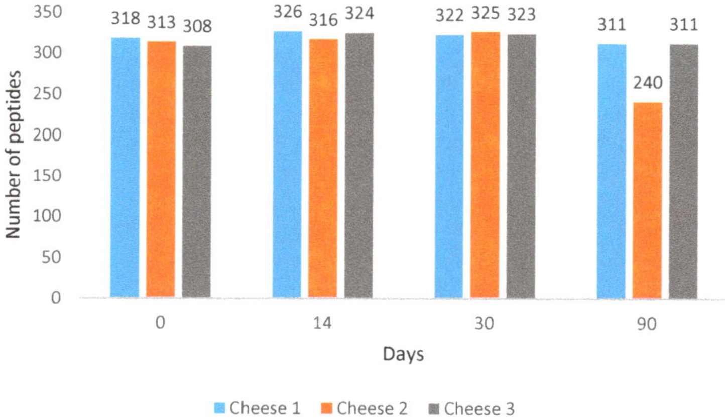

Figure 1. Number of identified peptides with unique amino acid sequences across 90 days of ripening.

Cheese 1 and Cheese 3 had 6.7- and 6-times higher summed peptide intensities compared to Cheese 2 (Figure S1).

At the same time, the average length of peptides in Cheese 2 was found to be longer than in other cheeses (Figure S2). All cheeses were subjected to comparative analysis to identify unique peptides at each measured point during cheese ripening. Results of the comparison are illustrated in Figure 2 that displays four Venn diagrams [21] for different days of ripening.

It was found that identified peptides during the first month of ripening were highly similar and accounted for approximately 93% identified peptides in all samples. During the ripening process, similarities in peptide composition between cheeses started to decrease. At the 90th day of ripening, Cheese 1 and Cheese 3 were more similar in peptide composition compared to Cheese 2, including over 60 identified peptides that were not present in Cheese 2.

Figure 3 indicates the difference in peptide accumulation and degradation pattern between days 0 and 90 for all three cheeses. Cheese 2, unlike the other two, displays the prevalence of peptide degradation compared to accumulation. This pattern is also consistent with the summed peptide intensities of each sample.

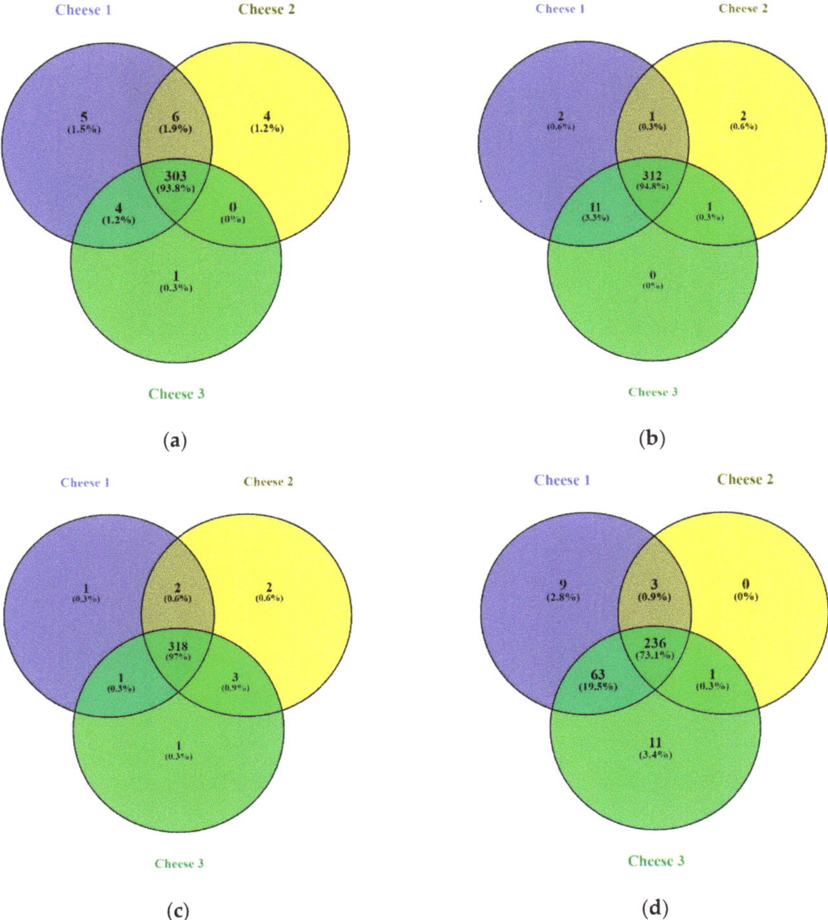

Figure 2. Venn diagrams of peptide distribution for Cheese 1, 2 and 3: (**a**) 0 days of ripening, (**b**) 14 days of ripening, (**c**) 30 days of ripening, (**d**) 90 days of ripening. Percentages in brackets denote proportion of all identified peptides across all cheeses.

For comparison between DDA and DIA-based approaches, the top 150 most abundant peptides per method were selected by their normalised intensities at the 90th day of ripening. It was found that 70 unique peptide sequences with a median length of nine amino acids were similar between DIA and DDA approaches, based on the identified peptides from Cheese 1. DIA-based approach results showed 80 unique peptide sequences with a median length of a peptide of seven amino acids. On the other hand, the DDA-based approach results showed 80 unique peptide sequences with a median length of 10 amino acids (Figure 4).

As for peptides identified in Cheese 2 sample, 58 unique peptides sequences (median length: 9 AA) were common between two methodologies, 92 unique peptide sequences (median length: 7 AA) belonged to DIA-based approach results and 92 unique peptides sequences (median length: 10 AA) were found only in the DDA-based approach results (Figure 5).

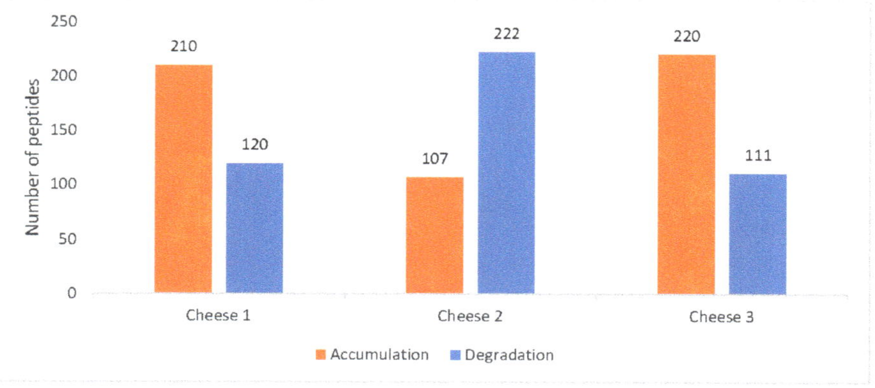

Figure 3. Cheese peptide profile trends between Days 0 and 90.

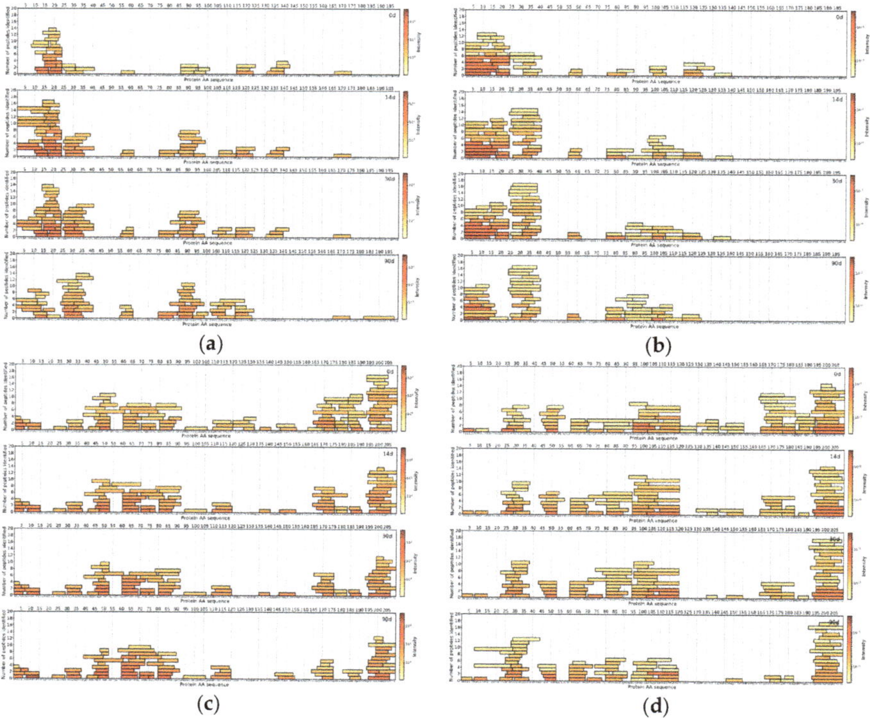

Figure 4. Cheese 1 α_{s1}-CN: (**a**) data-independent acquisition (DIA)-MS and (**b**) data-dependent acquisition (DDA)-MS. Cheese 1 β-CN: (**c**) DIA-MS and (**d**) DDA-MS. The *X*-axis represents the casein amino acid sequence, the *Y*-axis represents a number of peptides and colour represents the logarithmically scaled intensity of peptides.

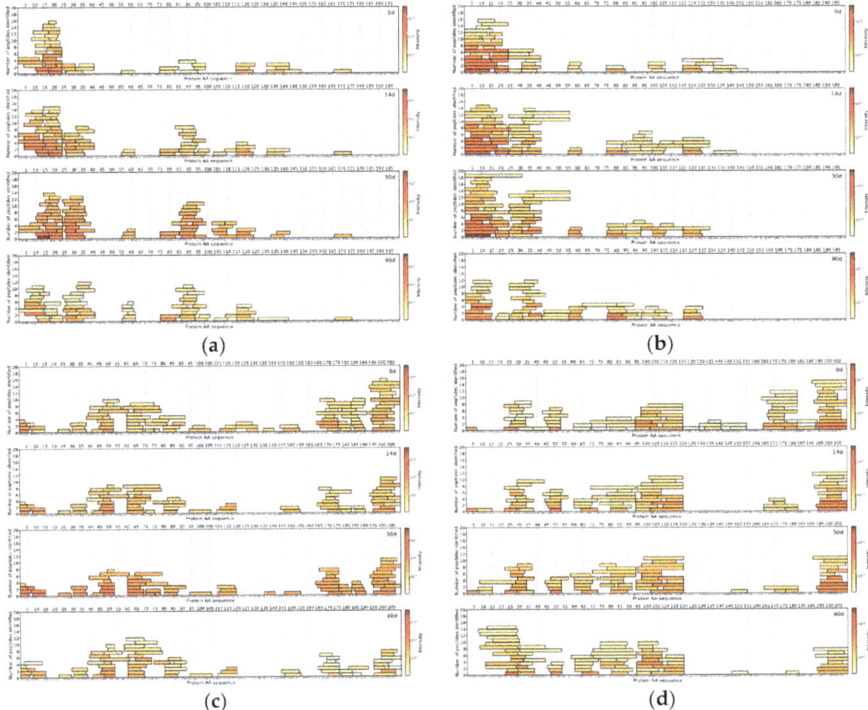

Figure 5. Cheese 2 α_{s1}-CN: (**a**) DIA-MS and (**b**) DDA-MS. Cheese 2 β-CN: (**c**) DIA-MS and (**d**) DDA-MS. The *X*-axis represents the casein amino acid sequence, the *Y*-axis represents a number of peptides and colour represents the logarithmically scaled intensity of peptides.

Peptides identified in Cheese 3 showed a similar trend as 74 unique peptide sequences (median length: 8 AA) were in common between two methodologies, 76 unique peptide sequences (median length: 7 AA) were found in DIA-based approach results and 76 unique peptide sequences (median length: 11 AA) belonged to the DDA-based approach (Figure 6).

Across all samples analysed with either DIA- or DDA-based approaches, peptides from α_{s1}-CN and β-CN comprised the majority of all detected peptides in the top 150 most abundant peptides.

In this study, we have also found several peptides, that have been previously reported to show bioactivity [6,22]: VPITPT (α_{s2}-CN f117-122), MPFPKYPVEPF (β-CN f109-119), EPVLGPVRGPFP (β-CN f195-206), DKIHPF (β-CN f47-52), YPFPGPIPN (β-CN f60-68), TPVVVPPFLQPE (β-CN f80-91), VPGEIVE (β-CN f8-14), VPSERYL (α_{s1}-CN f86-92), VLGPVRGPFP (β-CN f197-206) and YPFPGPI (β-CN f60-66).

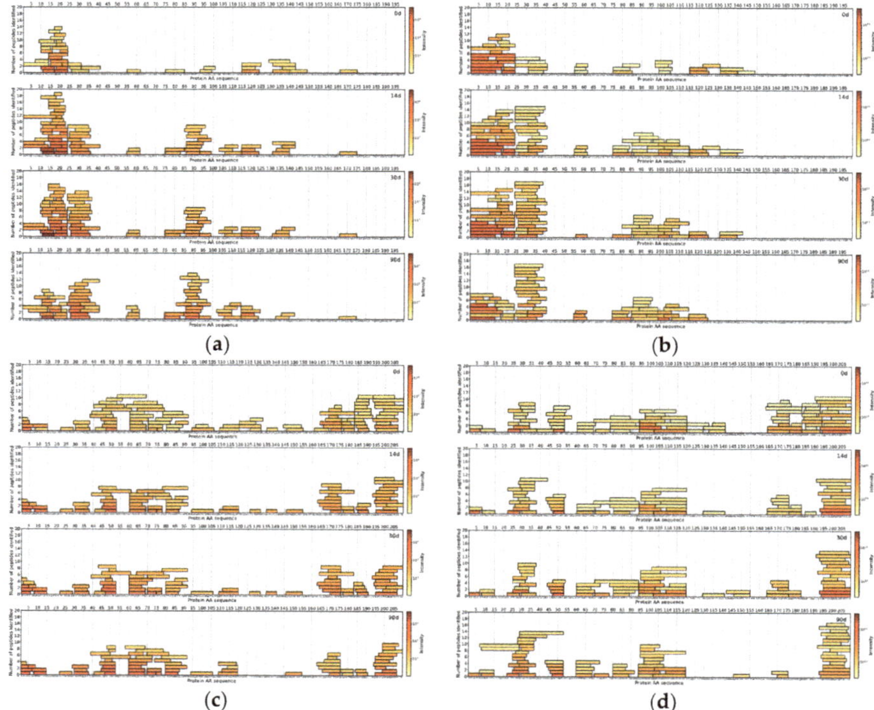

Figure 6. Cheese 3 α_{s1}-CN: (**a**) DIA-MS and (**b**) DDA-MS. Cheese 3 β-CN: (**c**) DIA-MS and (**d**) DDA-MS. The *X*-axis represents the casein amino acid sequence, the *Y*-axis represents a number of peptides and colour represents the logarithmically scaled intensity of peptides.

4. Discussion

During chromatographic method development, three peptide elution profiles were evaluated. The separation was performed with 40, 60 and 120 min analytical gradients to compare chromatographic performances and MS method compatibility. Figure 7 displays a base peak intensity chromatogram for a 40 min analytical gradient method. The narrowest extracted peak chromatogram was at 10 s at the base of the peak, providing a sufficient number of data points per peak (Figure S3). Therefore, as 60 and 120 min methods did not result in a higher number of identified peptides, a 40 min analytical gradient was selected as the one facilitating the best throughput.

Although the conventional DDA approach provides cleaner MS^2 spectra due to active isolation of the precursor, it suffers from a phenomenon known as data completeness problem [19]. While most abundant species get their corresponding MS^2 spectra recorded, less abundant species can potentially be missed. Due to sample-to-sample variation in analyte concentration in combination with precursor selection criteria even a peptide eluting at the same time might be missed out.

DIA is not only not subjected to the aforementioned drawback of DDA, but it also operates at a significantly higher acquisition speed due to minimising the time between MS^1 and MS^2 scan acquisition. Therefore, faster acquisition rate does not only allow to record data qualitatively, but also quantitatively. However, DIA-based methods with a quadrupole set for static transition exceedingly rely on chromatographic separation to minimise peptide coelution and hence, acquire cleaner MS^2 spectra.

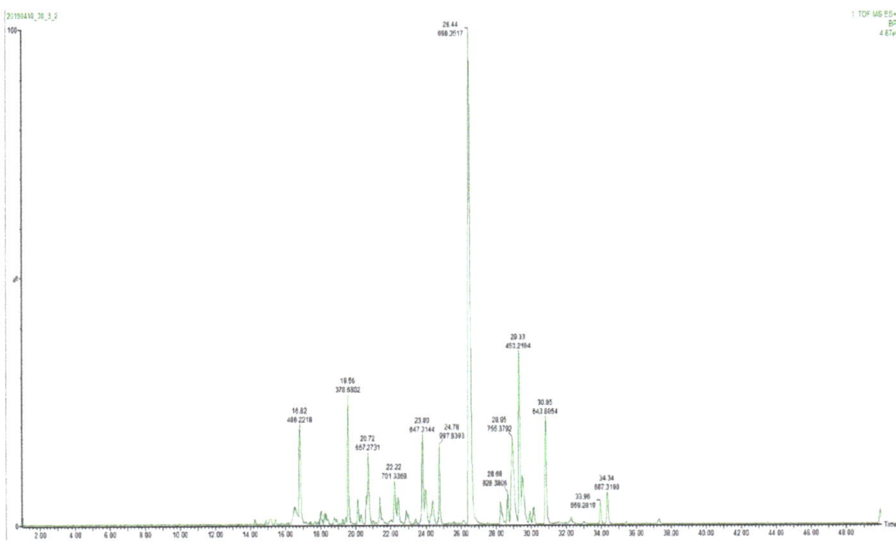

Figure 7. Base peak intensity (BPI) chromatogram of 40 min analytical gradient.

Furthermore, during synthesis in the mammary gland, caseins undergo post-translational changes in their primary structure [23]. One of the most important post-translational modifications in caseins is phosphorylation (at Ser, Thr and Tyr residues) and thus, analysis of phosphorylated peptides requires additional enrichment and purification step to decrease ion ionisation competition between non- and phosphorylated peptides [24]. With the current method, it is not possible to robustly analyse the peptides with every possible modification.

In recent years methods of further enhancement of a conventional DIA approach are gaining significant popularity. Active scanning (SONAR®, Waters and Scanning SWATH®) or stepped (SWATH®, Sciex) quadrupole and ion mobility separation (HDMS^E®, Waters and PASEF®, Bruker)-based DIA methods further expand the capabilities of DIA [25]. Active scanning or stepping quadrupole-based DIA methods significantly improve spectral clarity of MS^2 spectra by allowing fragmentation of only the ions confining within a quadrupole transmission profile. However, it loses a portion of the beam not confining to a quadrupole transmission window and hence, results in decreased overall sensitivity. Ion mobility separation based DIA, on the other hand, operates under a principle of preion mobility separation ion accumulation and subsequent release and hence, does not suffer from the ion loss of the quadrupole-based methods. As fragments can only exist when a precursor is present and fragments are inheriting the same drift time as the precursor due to the fact that mobility separation takes place before the fragmentation, it has been reported that ion mobility separation achieves a similar type of MS^2 clarity using the alignment of drift times and chromatographic profile of a precursor and fragments (HDMS^E®/PASEF®) [26,27]. Implementation of enhanced DIA methods would allow for even faster gradients and is worth further investigation.

A cut-off filter (3 kDa) was selected for sample preparation due to the unclear interaction of shorter cheese peptides with reversed-phase solid-phase extraction. In our work, we observed bias towards shorter peptides compared to Taivosalo et al. [14]. This bias could have been caused by either a natural bias of the given system towards shorter peptides, a decreased loss of shorter peptides due to not using of reverse-phase solid-phase extraction, or an increased loss of longer peptides due to the use of 3 kDa cut-off filter. As the overall number of peptides identified was lower than anticipated the use of

3 kDa cut-off filter should be further reviewed for its performance against conventional reversed-phase solid-phase extraction methods.

5. Conclusions

A rapid method was developed and successfully applied to the cheese peptidomics studies. The study allowed to indicate differences in cheese ripening caused due to the use of different starter cultures. Further methodology development is possible via the deployment of enhanced DIA approaches. Enhanced transmission of shorter peptides presents additional interest for future studies due to recorded bioactivity and sensory effects.

Supplementary Materials: The following are available online at http://www.mdpi.com/2304-8158/9/8/979/s1, Figure S1: Summed peptide intensities of Cheese 1, 2 and 3 at the day 90, Figure S2: Peptide length distribution across 3 cheeses at the day 90, Figure S3: Overlay of Base Peak Intensity and Extracted Ion Chromatogram for narrowest peak corresponding to an identified peptide.

Author Contributions: Conceptualisation, G.A., A.T. and S.J.; methodology, G.A.; software, G.A. and S.V.; instrumental setup Z.D.; data curation, S.J. and R.R.; writing—original draft preparation, G.A.; writing—review and editing, A.T., D.P., S.J. and R.V.; visualisation, D.P. and T.L.; project administration, G.A.; funding acquisition, R.V. All authors have read and agreed to the published version of the manuscript.

Funding: This work was partially supported by "TUT Institutional Development Program for 2016–2022" Graduate School in Biomedicine and Biotechnology receiving funding from the European Regional Development Fund under program ASTRA 2014–2020.4.01.16-0032 in Estonia.

Acknowledgments: Authors would like to acknowledge Tiina Kriščiunaite for input in data interpretation.

Conflicts of Interest: The authors declare no conflicts of interest.

References

1. Fox, P.F.; McSweeney, P.L.H. Proteolysis in cheese during ripening. *Food Rev. Int.* **1996**, *12*, 457–509. [CrossRef]
2. Fontenele, M.A.; Bastos, M.S.R.; dos Santos, K.M.O.; Bemquerer, M.P.; do Egito, A.S. Peptide profile of Coalho cheese: A contribution for Protected Designation of Origin (PDO). *Food Chem.* **2017**, *219*, 382–390. [CrossRef]
3. Galli, B.D.; Baptista, D.P.; Cavalheiro, F.G.; Negrão, F.; Eberlin, M.N.; Gigante, M.L. Peptide profile of Camembert-type cheese: Effect of heat treatment and adjunct culture Lactobacillus rhamnosus GG. *Food Res. Int.* **2019**, *123*, 393–402. [CrossRef]
4. Sforza, S.; Cavatorta, V.; Lambertini, F.; Galaverna, G.; Dossena, A.; Marchelli, R. Cheese peptidomics: A detailed study on the evolution of the oligopeptide fraction in Parmigiano-Reggiano cheese from curd to 24 months of aging. *J. Dairy Sci.* **2012**, *95*, 3514–3526. [CrossRef]
5. Sánchez-Rivera, L.; Martínez-Maqueda, D.; Cruz-Huerta, E.; Miralles, B.; Recio, I. Peptidomics for discovery, bioavailability and monitoring of dairy bioactive peptides. *Food Res. Int.* **2014**, *63*, 170–181. [CrossRef]
6. Nielsen, S.D.; Beverly, R.L.; Qu, Y.; Dallas, D.C. Milk bioactive peptide database: A comprehensive database of milk protein-derived bioactive peptides and novel visualization. *Food Chem.* **2017**, *232*, 673–682. [CrossRef]
7. Ardö, Y.; Lilbæk, H.; Kristiansen, K.R.; Zakora, M.; Otte, J. Identification of large phosphopeptides from β-casein that characteristically accumulate during ripening of the semi-hard cheese Herrgård. *Int. Dairy J.* **2007**, *17*, 513–524. [CrossRef]
8. Karametsi, K.; Kokkinidou, S.; Ronningen, I.; Peterson, D.G. Identification of Bitter Peptides in Aged Cheddar Cheese. *J. Agric. Food Chem.* **2014**, *62*, 8034–8041. [CrossRef] [PubMed]
9. Fallico, V.; McSweeney, P.L.H.; Horne, J.; Pediliggieri, C.; Hannon, J.A.; Carpino, S.; Licitra, G. Evaluation of Bitterness in Ragusano Cheese. *J. Dairy Sci.* **2005**, *88*, 1288–1300. [CrossRef]
10. Baptista, D.P.; Galli, B.D.; Cavalheiro, F.G.; Negrão, F.; Eberlin, M.N.; Gigante, M.L. Lactobacillus helveticus LH-B02 favours the release of bioactive peptide during Prato cheese ripening. *Int. Dairy J.* **2018**, *87*, 75–83. [CrossRef]
11. Reale, A.; Ianniello, R.G.; Ciocia, F.; Di Renzo, T.; Boscaino, F.; Ricciardi, A.; Coppola, R.; Parente, E.; Zotta, T.; McSweeney, P.L.H. Effect of respirative and catalase-positive Lactobacillus casei adjuncts on the production and quality of Cheddar-type cheese. *Int. Dairy J.* **2016**, *63*, 78–87. [CrossRef]

12. Gagnaire, V.; Mollé, D.; Herrouin, M.; Léonil, J. Peptides Identified during Emmental Cheese Ripening: Origin and Proteolytic Systems Involved. *J. Agric. Food Chem.* **2001**, *49*, 4402–4413. [CrossRef] [PubMed]

13. Sforza, S.; Ferroni, L.; Galaverna, G.; Dossena, A.; Marchelli, R. Extraction, Semi-Quantification, and Fast On-line Identification of Oligopeptides in Grana Padano Cheese by HPLC–MS. *J. Agric. Food Chem.* **2003**, *51*, 2130–2135. [CrossRef] [PubMed]

14. Taivosalo, A.; Kriščiunaite, T.; Seiman, A.; Part, N.; Stulova, I.; Vilu, R. Comprehensive analysis of proteolysis during 8 months of ripening of high-cooked Old Saare cheese. *J. Dairy Sci.* **2018**, *101*, 944–967. [CrossRef]

15. Baptista, D.P.; Araújo, F.D.S.; Eberlin, M.N.; Gigante, M.L. A Survey of the Peptide Profile in Prato Cheese as Measured by MALDI-MS and Capillary Electrophoresis: Peptide profile of Prato cheese. *J. Food Sci.* **2017**, *82*, 386–393. [CrossRef]

16. Rehn, U.; Petersen, M.A.; Saedén, K.H.; Ardö, Y. Ripening of extra-hard cheese made with mesophilic DL-starter. *Int. Dairy J.* **2010**, *20*, 844–851. [CrossRef]

17. Toelstede, S.; Hofmann, T. Sensomics Mapping and Identification of the Key Bitter Metabolites in Gouda Cheese. *J. Agric. Food Chem.* **2008**, *56*, 2795–2804. [CrossRef]

18. Yilmaz, M.T.; Kesmen, Z.; Baykal, B.; Sagdic, O.; Kulen, O.; Kacar, O.; Yetim, H.; Baykal, A.T. A novel method to differentiate bovine and porcine gelatins in food products: NanoUPLC-ESI-Q-TOF-MSE based data independent acquisition technique to detect marker peptides in gelatin. *Food Chem.* **2013**, *141*, 2450–2458. [CrossRef]

19. Hernández-Mesa, M.; Escourrou, A.; Monteau, F.; Le Bizec, B.; Dervilly-Pinel, G. Current applications and perspectives of ion mobility spectrometry to answer chemical food safety issues. *Trends Anal. Chem.* **2017**, *94*, 39–53. [CrossRef]

20. The UniProt Consortium. The Universal Protein Resource (UniProt). *Nucleic Acids Res.* **2008**, *36*, D190–D195. [CrossRef]

21. Oliveros, J.C.; Venny. An Interactive Tool for Comparing Lists with Venn's Diagrams. 2007–2015. Available online: https://bioinfogp.cnb.csic.es/tools/venny/index.html (accessed on 15 December 2019).

22. Sebald, K.; Dunkel, A.; Hofmann, T. Mapping Taste-Relevant Food Peptidomes by Means of Sequential Window Acquisition of All Theoretical Fragment Ion–Mass Spectrometry. *J. Agric. Food Chem.* **2019**. [CrossRef] [PubMed]

23. Farrell, H.M.; Jimenez-Flores, R.; Bleck, G.T.; Brown, E.M.; Butler, J.E.; Creamer, L.K.; Hickss, C.L.; Hollar, C.M.; NG-Kwai-Hang, K.F.; Swaisgood, H.E. Nomenclature of the proteins of cows' milk–sixth revision. *J. Dairy Sci.* **2004**, *87*, 1641–1674. [CrossRef]

24. Larsen, M.; Thingholm, T.; Jensen, O.; Roepstorff, P.; Jorgensen, T. Highly Selective Enrichment of Phosphorylated Peptides from Peptide Mixtures Using Titanium Dioxide Microcolumns. *Mol. Cell Proteomics* **2005**, *4*, 873–886. [CrossRef]

25. Ludwig, C.; Gillet, L.; Rosenberger, G.; Amon, S.; Collins, B.C.; Aebersold, R. Data-independent acquisition-based SWATH-MS for quantitative proteomics: A tutorial. *Mol. Syst. Biol.* **2018**, *14*, e8126. [CrossRef] [PubMed]

26. Alves, T.O.; D'Almeida, C.T.S.; Victorio, V.C.M.; Souza, G.H.M.F.; Cameron, L.C.; Ferreira, M.S.L. Immunogenic and allergenic profile of wheat flours from different technological qualities revealed by ion mobility mass spectrometry. *J. Food Compos. Anal.* **2018**, *73*, 67–75. [CrossRef]

27. Jeanne Dit Fouque, K.; Fernandez-Lima, F. Recent advances in biological separations using trapped ion mobility spectrometry–mass spectrometry. *Trends Anal. Chem.* **2019**, *116*, 308–315. [CrossRef]

© 2020 by the authors. Licensee MDPI, Basel, Switzerland. This article is an open access article distributed under the terms and conditions of the Creative Commons Attribution (CC BY) license (http://creativecommons.org/licenses/by/4.0/).

Article

Impact of Extending Hard-Cheese Ripening: A Multiparameter Characterization of Parmigiano Reggiano Cheese Ripened up to 50 Months

Paolo D'Incecco *, Sara Limbo, John Hogenboom, Veronica Rosi, Serena Gobbi and Luisa Pellegrino

Department of Food, Environmental and Nutritional Sciences (DeFENS), Università degli Studi di Milano, 20133 Milan, Italy; sara.limbo@unimi.it (S.L.); john.hogenboom@unimi.it (J.H.); veronica.rosi@unimi.it (V.R.); serena.gobbi@unimi.it (S.G.); luisa.pellegrino@unimi.it (L.P.)
* Correspondence: paolo.dincecco@unimi.it

Received: 31 January 2020; Accepted: 26 February 2020; Published: 2 March 2020

Abstract: Extending ripening of hard cheeses well beyond the traditional ripening period is becoming increasingly popular, although little is known about the actual evolution of their characteristics. The present work aimed at investigating selected traits of Parmigiano Reggiano cheese ripened for 12, 18, 24, 30, 40 and 50 months. Two cheeses per each ripening period were sampled. Although moisture constantly decreased and was close to 25% in 50-month cheeses, with a parallel increase in cheese hardness, several biochemical changes occurred involving the activity of both native and microbial enzymes. Capillary electrophoresis demonstrated degradation of α_{s1}- and β-casein, indicating residual activity of both chymosin and plasmin. Similarly, continuous release of free amino acids supported the activity of peptidases deriving from lysed bacterial cells. Volatile flavor compounds, such as short-chain fatty acids and some derived ketones, alcohols and esters, evaluated by gas chromatography with solid-phase micro-extraction, accumulated as well. Cheese microstructure was characterized by free fat trapped in irregularly shaped areas within a protein network, with native fat globules being no longer visible. This study showed for the first time that numerous biochemical and structural variations still occur in a hard cheese at up to 50 months of aging, proving that the ripening extension deserves to be highlighted to the consumer and may justify a premium price.

Keywords: cheese ripening; ripening extension; cheese microstructure; free amino acids; capillary electrophoresis; proteolysis; volatile compounds; confocal laser scanning microscopy

1. Introduction

Parmigiano Reggiano is an Italian extra-hard cheese made from raw milk. Being a protected designation of origin (PDO) product, it is produced in a restricted geographical area using the traditional cheesemaking described in the product specification [1]. Milk is partly skimmed by natural creaming, poured into the traditional open copper vat with natural whey starter and calf rennet in order to achieve coagulation at 33–34 °C within 8–10 min. The curd is cut into rice-grain-sized granules while the temperature is progressively raised up to 53–54 °C in 10–12 min under gentle stirring. When stirring is stopped, the curd granules sediment and aggregate at the bottom of the vat due to the compression that hot whey exerts for about an hour. The curd is then extracted from the vat, cut into two portions that are set into circular molds and kept there to slowly cool down, acidify and lose more whey for about two days. Afterward, cheese wheels (~35 kg each) are salted in brine for 20–22 days and then ripened in a ripening room at about 18 °C and 80% relative humidity. The minimum ripening period for PDO Parmigiano Reggiano is 12 months.

It is commonly recognized that quality of artisanal cheeses mostly relies on the skill of the cheesemaker to master the vat process by day-to-day adapting the extent and timing of the different actions to the characteristics of the raw materials, namely raw milk and the natural starter culture. Nevertheless, the typical cheese characteristics develop during the subsequent ripening period, when a variety of profound and complex changes take place. These changes imply phenomena that are particularly interesting in long-ripened cheeses made from raw milk, such as Parmigiano Reggiano and Grana Padano, and that have been addressed in many research studies and review articles [2,3].

Enzymes such as the native plasmin and the rennet chymosin, both largely retained in the curd, soon attack casein fractions, initiating the proteolytic maturation of cheese. Further degradation of casein into peptides [4] and free amino acids [5] takes longer and needs the contribution of microbial peptidases. Microbial lipases are responsible for the release of free fatty acids that contribute to cheese flavor development [6]. Both proteolysis and lipolysis products proved to reach characteristic profiles in raw milk PDO cheeses at different ages [6,7], confirming that the microbial populations are rather homogeneous within the respective production areas and act in a repeatable way because the selective action of both the cheesemaking and ripening conditions are markedly consistent among cheese factories.

A variety of microbial populations are natural contaminants of raw milk, including the so-called nonstarter lactic acid bacteria (NSLAB), whereas others are added with the natural starter, predominated by lactic acid bacteria. Different populations are active or inactive throughout ripening, depending on the selectivity of environmental conditions that are progressively changing within the cheese. When starter lactic acid bacteria (SLAB) decline, NSLAB rise and largely contribute to cheese flavor development, as mesophilic facultatively heterofermentative species dominate in this population [2,8].

During the recent years, increasing amounts of hard cheeses such as Parmigiano Reggiano, Grana Padano, Comté, Cheddar or Gouda, are kept ripening for longer periods than usual in the past. Considering that consumption rates are globally increasing for those cheeses and new market areas are presently accessible, it is of interest to the manufacturers to provide the consumers with a wider array of products, including cheeses having more pronounced taste and brittle texture. To date, there has been little research focus on cheese ripening periods over 24 months [5,9].

In this study, we tracked various composition, structure and appearance characteristics of Parmigiano Reggiano cheeses during a ripening period up to 50 months, in order to elucidate whether some of these are still evolving over such a long period and might be useful in characterizing the longer-aged cheeses. Regardless of the distinctive sensory properties that are well known and appreciated by the consumers worldwide [10], we were interested in estimating the significance of extending the ripening of Parmigiano Reggiano cheese far more than the usual 12–18 months. Therefore, different aspects were addressed and a wide array of analytical parameters were tested on cheeses ripened from 12 to 50 months.

2. Materials and Methods

2.1. Cheese Samples

A total of 12 Parmigiano Reggiano (PR) cheeses were obtained from six factories located in the production area of the PDO Parmigiano Reggiano cheese. Two cheeses of a specific ripening period were obtained per factory. The manufacture and ripening conditions provided by PDO Parmigiano Reggiano specification and described in Figure S1 were followed. Cheese ages were 12 (12mo), 18 (18mo), 24 (24mo), 30 (30mo), 40 (40mo) and 50 months (50mo) with ±1 month of incertitude. From each cheese wheel, two vertical slices (around 1 kg each) were taken. One slice was used to count white spots and tyrosine crystals visible on the surface (100 cm^2) as described by D'Incecco et al. [11] (Figure 1) and then used for color, rheology and microscopy analyses that were immediately conducted in sequence in order to avoid cheese drying. The other slice was finely grated using a domestic grinder, after removal of 0.5-cm rind layer, and submitted to chemical analysis within the next day.

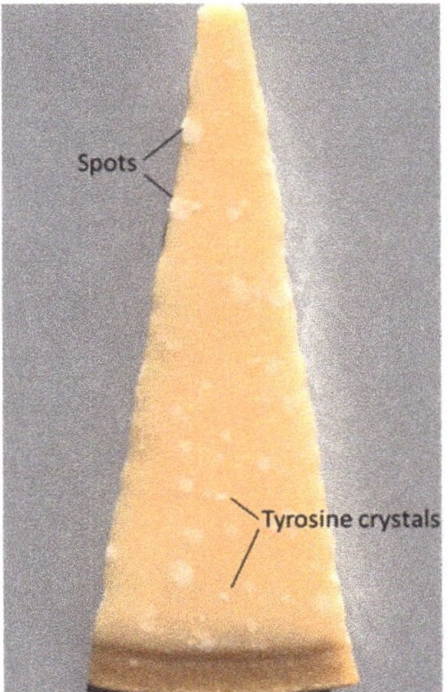

Figure 1. Slice of 40-month ripened Parmigiano Reggiano cheese.

2.2. Cheese Composition Analysis

Cheese samples were analyzed for moisture, fat and protein content using the International Standard methods of the International Dairy Federation, as described previously [7]. Protein content was calculated using 6.38 as conversion factor. The moisture in the nonfat substance (MNFS) was calculated as 100 × moisture content/(100−fat content). Analyses were carried out in duplicate.

2.3. Proteolysis Extent

2.3.1. Casein and Peptides

Intact casein fractions and major peptides were analyzed by capillary zone electrophoresis (CZE), adopting the conditions described by D'Incecco et al. [12]. Grated cheese (1 g) was dispersed in a sample buffer (10 mL) and kept at room temperature for at least 4 h. Sample buffer was prepared by adding 200 mL of 60% (*w/v*) urea solution in Millipore MilliQ purified water and 300 mL of urea 60% (*w/v*)/methylhydroxyethylcellulose (MECH) 0.15% (*w/v*) in MilliQ water with 7.44 g of ethylenediaminetetraacetic acid disodium salt dihydrate, 6.06 g of tris(hydroxymethyl)aminomethane, 2.64 g of 3-(N-morpholino)propanesulfonic acid and 0.77 g of dithiothreitol. The sample was further diluted 1:5 with the same buffer, filtered with 0.22 μm polyvinylidene fluoride membrane filter (Millipore) and separated using a hydrophilically coated capillary column (50 μm i.d., 0.05 μm coating, 500 mm effective length, 100 × 800 μm slit opening, DB-WAX 126-7012, J&W Agilent Technologies, Milan, Italy). Separation was carried out at 45 °C with linear gradient from 0 to 30 kV in 4 min, followed by constant voltage at 30 kV for 36 min, using P/ACE™ MDQplus capillary electrophoresis equipment (AB Sciex, Milan, Italy) including a UV detector set at 214 nm. Separation buffer was prepared as follows: 60 mL of urea 60% (*w/v*)/MECH 0.15% (*w/v*) in MilliQ water solution were added with 4.38 g of citric acid monohydrate, 0.59 g of trisodium citrate dehydrate and 40 mL of MilliQ water. Separation buffer was filtered with 0.45 μm regenerated cellulose membrane filter (Agilent Technologies). Peak identification in the obtained electropherograms is shown in Figure 2. The peak area ratios between selected casein or peptide fractions were calculated, considering normalized peak area (peak area counts/migration time), as follows:

$$\frac{\alpha_{s1} - I}{\alpha_{s1}} = \frac{\alpha_{s1} - CN - I\ 8P}{\alpha_{s1} - CN\ 8P} \tag{1}$$

$$\frac{\alpha_{s1}f(1-23)}{\alpha_{s1}} = \frac{\alpha_{s1}-CN\ f(1-23)}{\alpha_{s1}-CN\ 8P} \tag{2}$$

$$\frac{\Sigma\gamma}{\Sigma\beta} = \frac{\gamma1-CN\ A1 + \gamma1-CN\ A2 + \gamma2-CN\ A + \gamma3-CN\ A + p\gamma3-CN\ A}{\beta-CN\ B + \beta-CN\ A1 + \beta-CN\ A2} \tag{3}$$

where α_{s1}-CN 8P is α_{s1}-CN with eight phosphorylated serine residues.

Cheese age, expressed in months, was calculated according to Masotti et al. [5] as follows:

$$\text{Cheese age} = 0.91\left(100\frac{p\gamma_3-CN\ A}{p\gamma_3-CN\ A + \gamma_3-CN\ A}\right)+4.33 \tag{4}$$

where $p\gamma_3$-CN is the pyroglutamyl-γ3-casein.

Analyses were carried out in duplicate.

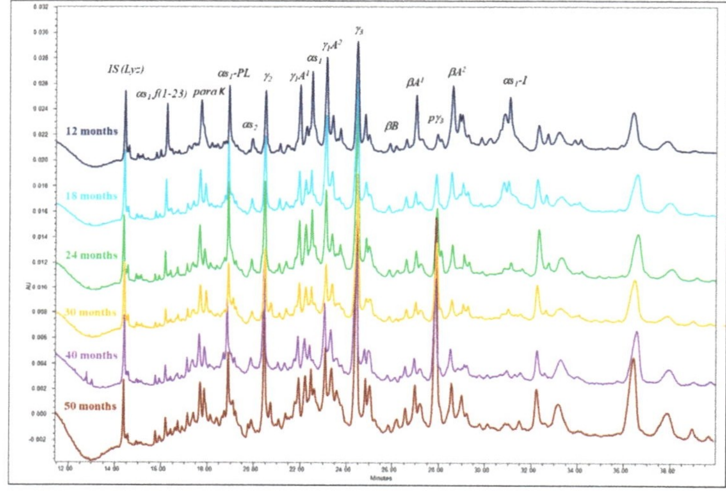

Figure 2. Capillary electrophoresis patterns of Parmigiano Reggiano cheeses ripened up to 50 months.

2.3.2. Free Amino Acids

Ion-exchange chromatography was used for free amino acid (FAA) analysis following the conditions described by Hogenboom et al. [13]. The equipment was an amino acid analyzer Biochrom 30plus (Biochrom Ltd., Cambridge, UK). Free amino acids were extracted from grated cheese (1.5 g) previously dissolved in 0.2 N sodium citrate buffer, homogenized then deproteinated with 7.5% 5-sulfosalicylic acid. The obtained solution (10 mL) was added with 2 mL of a 600 mg/L solution of Norleucine as an internal standard and diluted in 0.2 N lithium citrate. Extracts were filtered on 0.2 μm cellulose acetate filter (Millipore) prior to injection. The chromatographic conditions were those recommended by the manufacturer. A multipoint calibration was used for quantitation of 21 amino acids. Analyses were carried out in duplicate.

2.4. Volatile Compounds

The determination of volatiles was carried out on the samples with different ages. Five grams of grated cheese was weighted in a 20 mL glass vial sealed with an aluminum cap provided with a silicon septum (HTA, Brescia, Italy). A carboxen-polydimethylsiloxane-divinylbenzene (CAR-PDMS-DVB; 50/30 μm × 1 cm) (Supelco, Bellefonte, PA, USA) was used to collect volatiles from the samples using an automatic SPME autosampler (HTA, Brescia, Italy) set at the following conditions: incubation for 10 min at 40 °C; agitation for 5 min; extraction for 45 min; desorption for 20 min. After the extraction step, the volatiles were release in the injector of a gas chromatograph (Perkin Elmer Autosystem XL Gas Chromatograph) coupled with a mass spectrometer (Turbomass, Perkin Elmer, Italy). The injector was set at 250 °C and the injection mode was splitless for 0.50 min. The gas-chromatographic separation was carried out with a Stabilwax-MS column (30 m × 0.250 mm × 0.25 µm; Restek, Bellefonte, PA, USA) using helium as carrier at flow rate of 1.2 mL/min. The oven temperature was initially set at 40 °C and held for 8 min, ramped at 4 °C/min up to 220 °C and held for 15 min. The transfer line temperature was set at 200 °C and the source temperature at 250 °C. The mass spectrometer operated in electron ionization mode at 70 eV using the full scan mode. The MS detector registered the m/z in the range from 35 up to 350 Da. The ions used for identification were chosen according to the National Institute of Standards and Technology (NIST) MS Search 2.0 library and validated by external standard comparisons of ion fragmentation patterns and by calculating the linear retention index (LRI) running an alkane standard solution (C8–C20, Merck, Italy). Values were expressed as area units/10,000. Triplicate injections were carried out for each cheese sample.

2.5. Cheese Color

Color analysis was carried out on cheese slice surface using a portable tristimulus colorimeter (Minolta Chroma Meter CR 300—Minolta, Osaka, Japan), equipped with an 8 mm viewing port, illuminant C source and standard observer. Color coordinates (L^*, a^*, b^*) were measured in triplicate on each cheese slice and different color indexes were calculated [14]. Yellowness index (YI) [15] was calculated as:

$$\mathrm{YI} = 142.86 \frac{b^*}{L^*} \tag{5}$$

Total color difference (ΔE^*) between the color index recorded at each ripening time and that recorded at 12 months was obtained from the equation:

$$\Delta E^* = (\Delta a^{*2} + \Delta b^{*2} + \Delta L^{*2})^{1/2} \tag{6}$$

Hue angle (h^*) was calculated as:

$$h^* = \tan^{-1}\left(\frac{b^*}{a^*}\right) \tag{7}$$

2.6. Cheese Texture Analysis

Texture was analyzed using a TA-XT Plus (Stable Micro System, Surrey, England) texture analyzer equipped with the fracture wedge set comprising upper and lower wedges with cutting angle of 30° and 30 mm width. The upper wedge was connected directly to the load cell. The wedge fracture test was carried out on three cheese portions (2 cm × 2 cm × 2 cm) that were cut from each cheese slice at 5-cm depth below the rind of the round side and half-height of the cheese and conditioned at room temperature before analysis. Penetration was performed at a constant crosshead speed of 1 mm/s until either 70% of height or the fracture of the cheese cube was reached. The force/time curves were used to calculate cheese hardness (N) and the fracture deformation or brittleness (mm) according to manufacturer guidelines.

2.7. Cheese Microstructure

Microstructure of cheese samples was analyzed by Confocal Laser Scanning Microscopy (CLSM) as described previously [16]. Three portions of cheese (around 2 mm × 2 mm ×1 mm) were taken from each cheese slice at 5-cm depth using a razor blade. Samples were stained with Nile Red (Sigma Aldrich, St Louis, MO, USA) and Fast Green FCF (Sigma Aldrich) to visualize fat and protein matrix, respectively. The staining was carried out as follows: the stock solutions of Nile Red (1 mg/mL in dimethyl sulfoxide) and Fast Green (1 mg/mL in Millipore MilliQ purified water) were diluted tenfold just prior to 5-min staining. Samples were analyzed by using an inverted confocal laser scanning microscope A1+ (Nikon,

Minato, Japan). The excitation/emission wavelengths were set at 488 nm/520–590 nm for Nile Red and at 638 nm/660–740 nm for Fast Green FCF [17]. Images are presented as maximum projection of 23 optical sections stacked together with separation between layers set at 0.30 μm. Image analysis was performed using Vision4D software (Arivis, AG, Germany) on maximum projection of CLSM z-stack images. Porosity was calculated as the ratio between the nonfluorescent volume (μm^3) and the total fluorescent volume (μm^3).

2.8. Statistical Analysis

The data of composition, FAA, structure, rheology and color were assessed by one-way analysis of variance (ANOVA) and significant differences were considered at $p < 0.05$ level as evaluated by Tukey's test using SPSS Win 12.0 program Version 22 (SPSS Inc. IBM Corp., Chicago, IL). Principal component analysis (PCA) was carried out on selected variables showing a specific trend with respect to ripening time. Before PCA, data were preprocessed using the auto-scale mode and transformed using the normalized method. The Unscrambler v.9.7 software (Camo Software AS, 2007, Oslo, Norway) was used. Differences at $p < 0.05$ (*); $p < 0.01$ (**) and $p < 0.001$ (***) were considered significant.

3. Results and Discussion

3.1. Composition Analysis

The gross composition of the PR cheese samples of different ages is shown in Table 1. As expected, moisture was the most intensively changing among cheese components. Our data showed a progressive decrease during the considered period, although the variability of mean values was quite high in some cases. Values as low as 25 g/100 g cheese were reached in the 50mo ripened cheeses, which were markedly lower than the moisture content of 27 g/100 g measured by Malacarne et al. [18] in a 54-month old PR cheese. Fox et al. [19] attributed the maintenance of Parmesan cheese quality to its low levels of both moisture content (29.2%) and water activity (0.917). The protein content of the cheeses roughly varied from 31.2 and 33.8 g/100 g, with no characteristic trend over the ripening time. Concomitantly, the fat content was unusually variable (29.2–35.7 g/100 g) and the range of values on dry matter basis was wide as well (42.1–47.9 g/100 g). A high fat content makes cheese structure softer, especially when the moisture content is low [20,21]. Likely, when making cheeses destined to a prolonged ripening, the cheesemakers intentionally remove less fat from raw milk by shortening the creaming time in order to achieve this effect. To exclude the variability of fat content on cheese moisture, MNFS can be considered. Indeed, this parameter showed that the proportion of moisture decreased throughout the whole ripening period. Cheeses with MNFS < 51% are classified as extra-hard by Codex Alimentarius [22].

Table 1. Gross composition (g/100 g) of Parmigiano Reggiano cheeses ripened from 12 to 50 months.

Ripening Time (Months)	Moisture	Protein	Fat	FDM[1]	MNFS[2]
12	31.72 [c] ± 0.07	31.17 [a] ± 0.07	30.78 [ab] ± 0.02	45.08 [b] ± 0.03	45.83 [c] ± 0.09
18	30.61 [bc] ± 2.42	33.72 [d] ± 0.15	29.23 [a] ± 2.33	42.08 [a] ± 1.87	43.20 [b] ± 2.02
24	29.39 [bc] ± 1.83	31.37 [ab] ± 0.32	32.61 [bc] ± 2.14	46.15 [b] ± 1.84	43.59 [bc] ± 1.32
30	28.86 [abc] ± 0.48	31.75 [bc] ± 0.16	32.48 [bc] ± 0.38	45.65 [b] ± 0.25	42.73 [b] ± 0.48
40	28.03 [ab] ± 0.31	33.79 [d] ± 0.11	30.43[ab] ± 0.30	42.29 [a] ± 0.26	40.30 [a] ± 0.29
50	25.34 [a] ± 0.52	31.93 [c] ± 0.66	35.74[c] ± 0.10	47.87 [b] ± 0.43	39.43 [a] ± 0.85

The results are expressed as the mean ± the standard deviation. [a,b,c,d] Mean values within a column with different superscripts are significantly different ($p < 0.05$; Tukey's test). [1] FDM = fat in dry matter. [2] MNFS = moisture in nonfat substance.

3.2. Proteolysis

The extent of primary proteolysis was evaluated by CZE as the extent of casein fraction degradation (Figure 2). The CZE patterns evidenced very extensive changes taking place in PR cheese during the whole prolonged ripening period considered in this study and, nevertheless, many peaks corresponding to both intact casein and large fragments were still present in the most aged samples. Besides the specific cleavage of k-CN, rennet chymosin typically cleaves α_{s1}-CN at Phe_{23}-Phe_{24}, splitting the protein chain into two fragments, i.e., α_{s1}-CN f(1-23) and f(24-199), also called α_{s1}-I-CN. A progressive decrease of both α_{s1}-CN and the main derived fragments occurred throughout the whole ripening period (Figure 2). The peak-area ratio between these two fragments and the intact α_{s1}-CN slightly decreased over time, indicating that the degradation of the former proceeded further (Table 2). To our knowledge, no direct evidence is available in the literature of persistence of chymosin activity in long-ripened cheeses. It has been shown that, in high-temperature cooked cheeses, chymosin is inactivated [23,24] and the main α_{s1}-CN-derived fragments accumulate during the first 4-6 months of ripening. In contrast to this current view, Hynes et al. [25] demonstrated that in laboratory-manufactured Reggianito Argentino cheese cooked at 52 °C, not at 60 °C, chymosin partly reactivated during ripening, and this recovered activity contributed to α_{s1}-CN degradation. The authors, however, prolonged their observation over 3 months only. Considering that conditions other than cooking temperature [23,24] also concur in chymosin retention in the curd and inactivation, it can be reasonably assumed that a low quantity of residual chymosin could proceed to slowly degrade casein, likely in a localized manner.

Table 2. Casein fraction ratios (αs_1-I/$\alpha s1$, $\alpha s1$ f(1-23)/ $\alpha s1$ and $\Sigma\gamma/\Sigma\beta$), calculated cheese age (months) and total free amino acids (FAA) (g/100 g protein) in Parmigiano Reggiano cheeses ripened from 12 to 50 months.

Ripening Time (Months)	αs_1-I/αs_1 (1)	αs_1 f(1-23)/αs_1 (2)	$\Sigma\gamma/\Sigma\beta$ (3)	Calculated Cheese Age (4)	FAA
12	0.7	0.4	2.7	13	24.2 ± 0.08 [a]
18	0.6	0.3	5.0	22	24.6 ± 0.17 [a]
24	0.3	0.2	5.4	25	28.4 ± 0.25 [b]
30	0.3	0.2	8.7	34	28.6 ± 1.30 [b]
40	0.4	0.3	10.0	39	28.4 ± 0.26 [b]
50	0.2	0.2	12.1	49	30.0 ± 0.86 [c]

The results are expressed as the mean ± standard deviation. [(1)(2)(3)(4)] Equations presented in Section 2.5. [a,b,c] Mean values within FAA column with different superscripts are significantly different ($p < 0.05$; Tukey's test).

Plasmin is reported to be a major proteolytic enzyme in cooked cheeses [24,26], with β-CN being its primary substrate. This last is progressively degraded into the Ɣ-CNs which, on the contrary, are rather stable and accumulate [4,23]. The ratio between Ɣ-CNs and β-CNs ($\Sigma\gamma/\Sigma\beta$) progressively increased by a factor of 4.5 in the ripening time from 12 to 50 months (Table 2), suggesting that plasmin activity proceeded. Consistent with these findings, Mayer et al. [26] showed that β-CN was no longer detectable by gel electrophoresis in a 48-month aged PR, while it was in less aged cheeses. In a previous study, we proposed the pyroglutamyl-$Ɣ_3$-CN (p-$Ɣ_3$-CN) as an indicator of cheese age, which proved to be accurate for both Grana Padano and PR cheeses [5]. This peptide originates from the cyclization of the N-terminal residue of glutamic acid of $Ɣ_3$-CN and proved to be highly stable to further proteolysis during cheese ripening. In the present study, we have evaluated the actual cheese age of the PR samples using this approach and found values pretty close to those declared by manufacturers (Table 2). This finding strongly supports the persistent plasmin activity, since $Ɣ_3$-CN increased progressively and therefore the precursor of p-$Ɣ_3$-CN was freely available.

Secondary proteolysis, typically operated by bacterial enzymes, is well described by free amino acids (FAA) since they are the final products of enzymatic splitting of peptones and peptides [23]. Surprisingly, the content of FAA still significantly increased in PR cheeses from 12 to 50 months, when utilization by LAB is expected to be over, suggesting that proteases released by lysed cells can still have a role (Table 2). However, the limited number and variable origin of the samples considered in this study did not allow to reveal differences in FAA content in cheeses with similar age. Nevertheless, our data indicated that, in such long-ripened cheeses, approximately one-third of the protein is present in a directly utilizable form for humans. To the knowledge of the authors, no cheese types other than PR and Grana Padano display such high FAA content.

As previously observed in other cheese types [7,19,20], glutamic acid (Glu) was by far the most abundant among FAA (Table S1). Besides being split from proteins and peptides, Glu may also derive from glutamine (Gln) deamidation by glutaminase and, in turn, can be decarboxylated to γ-aminobutiric acid (GABA) by glutamate decarboxylase. Both Gln deamidation and Glu decarboxylation are involved in acid-resistance mechanisms identified in selected LAB [27–29]. The involved enzymes are more abundant in the cytoplasm and therefore can be active after cell lysis [23]. Indeed, almost no Gln was detected in PR cheeses ripened for 30 months or longer. The same pattern was found in long-ripened Grana Padano cheese [5]. Accumulation of aspartic acid (Asp) was slower, compared to that of Glu, and a smaller concentration of Asp was found in the 50mo PR cheeses. Although scarcely documented, Asn deamidation activity to form Asp was observed in selected LAB starters, especially facultatively heterofermentative species [30].

3.3. Volatile Compounds

The composition of the volatile compounds (VOCs) of PR cheeses ripened from 12 up to 50 months and detected by SPME-GC-MS is compiled in Table 3, where the identified substances are listed. Short-chain odd-numbered free fatty acids (FFA), esters and ketones were the most abundant compounds and accounted for 86%, 7% and 6% of the total area, respectively.

The total volatile fraction roughly tripled across the considered ripening period, particularly because of the increased amount of short-chain FFA, as similarly observed by Malacarne et al. [6] in PR cheese during 24-month ripening. To the best of our knowledge, no studies investigated the relationship between the autolysis of mesophilic NSLAB and the levels of volatile compounds in cheese at such long ripening times. However, the presence of free fat in large areas embedded in the protein network, as was observed by CLSM in our cheese samples (see Section 3.4), is a condition that may largely favor enzyme activity towards triglycerides.

In fact, in extra-hard cheeses, an extensive lipolysis of triglycerides can occur, supported by both microbial and native milk enzymes like lipases and esterases, and the FFA produced can directly or indirectly contribute to cheese aroma development. Esterases responsible for hydrolyzing short acyl ester chains (C2–C8) are intracellular enzymes, and those from mesophilic nonstarter lactobacilli (NSLAB) are reported to be the main contributors to short-chain FFA accumulation during ripening of PR cheese [2]. Lipases hydrolyze longer acyl ester chains that are characterized by more than 10 carbons [31].

As shown in Table 3, butanoic, hexanoic and octanoic short-chain fatty acids represented the main volatile acids, and their contribution increased during ripening. Butanoic acid is responsible for buttery–cheesy flavor, hexanoic acid for sweaty and sometimes pungent flavor and octanoic acid for goat-like flavor [32]. Qian and Burbank [33] indicated odor activity values (OAVs, i.e., the ratio between the concentration and the flavor threshold) equal to 0.5, 28 and 320 for octanoic, hexanoic and butanoic acid, respectively, and together the three FFA mainly contributed to the typical aroma of PR cheese. Few branched FFA were identified, especially at the beginning of the ripening (18–24 months): 2-methyl butanoic acid and and, to a greater extent, 3-methyl butanoic acid. These branched FFA have similar aroma characteristics as the corresponding linear FFA, and they are characterized by a lower threshold because of their higher vapor pressure [33]; therefore, they can subtly contribute to the overall flavor of PR cheese.

Short- and medium-chain FFA themselves contribute to the aroma while other FFA are precursors for the formation of other flavor compounds, such as methyl ketones, lactones, esters, alkanes and secondary alcohols [23]. The presence of these secondary volatile compounds plays an important role in enhancing cheese flavor complexity [34]. In PR cheese, ethyl butanoate, ethyl hexanoate and ethyl octanoate were detected, and their presence tended to increase during ripening. These esters have strong fruity and floral notes [35] and are primarily responsible for the fruity aroma in PR cheese due to their very low sensory threshold (on the order of 0.04–0.6 mg/kg fat) and high OAV [33]. Most of these compounds were already recognized in hard raw-milk cheeses. A predominance of both FFA and ketones among VOCs was also found by Abbatangelo et al. [36] in PR cheese, by Lazzi et al. [37] in 13-month Grana Padano cheese and by Ceruti et al. [38] in Reggianito cheese.

Table 3. Volatile compounds (area units/10,000) identified in Parmigiano Reggiano cheeses ripened from 12 to 50 months.

Compound	LRI [a]	Identification [b]	Ripening Time (Months)					
			12	18	24	30	40	50
Acids								
Acetic acid	1455	MS, PI	1000 ± 200	850 ± 270	950 ± 60	1300 ± 120	1100 ± 110	1200 ± 90
Propanoic acid	1546	MS, PI	3 ± 0	24 ± 2	4 ± 2	7 ± 1	8 ± 1	6 ± 1
Butanoic acid	1640	MS, PI	4300 ± 300	4500 ± 700	5500 ± 300	8100 ± 500	9300 ± 400	11000 ± 890
Pentanoic acid	1755	MS, PI	18 ± 1	21 ± 4	41 ± 14	73 ± 8	97 ± 9	91 ± 7
Hexanoic acid	1870	MS, PI	4800 ± 900	5600 ± 760	7000 ± 700	12000 ± 1200	12000 ± 890	17000 ± 1300
Heptanoic acid	1957	MS, PI	ND	6 ± 3	28 ± 21	74 ± 50	23 ± 15	58 ± 23
Octanoic acid	2066	MS, PI	ND	71 ± 10	170 ± 60	1300 ± 1200	500 ± 67	1700 ± 500
2-Methylbutanoic acid		T	4 ± 0.3	4 ± 0.3	3 ± 0.5	2 ± 0.4	3 ± 0.2	3 ± 0.2
3-Methylbutanoic acid		T	40 ± 0.3	36 ± 5	16 ± 5	20 ± 3	6 ± 1	20 ± 3
Sum of areas			*10165*	*11112*	*13712*	*22876*	*23037*	*31078*
Ketones								
2-Pentanone	990	MS, PI, ST	150 ± 12	400 ± 90	400 ± 110	300 ± 70	300 ± 50	340 ± 50
2-Heptanone	1190	MS, PI, ST	200 ± 40	570 ± 180	700 ± 40	600 ± 130	600 ± 70	630 ± 120
2-Nonanone	1400	MS, PI, ST	80 ± 2	170 ± 40	400 ± 100	290 ± 100	270 ± 60	700 ± 200
8-Nonen-2-one	1495	MS, PI	ND	15 ± 2	30 ± 15	40 ± 4	20 ± 4	70 ± 21
Sum of areas			*430*	*1155*	*1530*	*1230*	*1190*	*1740*
Alcohols								
2-Pentanol	1172	MS, PI, ST	95 ± 8	97 ± 76	86 ± 83	45 ± 8	18 ± 4	48 ± 18
2-Heptanol	1320	MS, PI	45 ± 4	33 ± 23	38 ± 23	17 ± 2	17 ± 4	37 ± 17
Sum of areas			*140*	*130*	*124*	*62*	*35*	*85*
Esters								
Ethyl butanoate	1055	MS, PI	250 ± 1	440 ± 49	150 ± 14	580 ± 96	330 ± 96	500 ± 87
Ethyl hexanoate	1240	MS, PI	430 ± 54	1000 ± 170	230 ± 70	2000 ± 75	720 ± 132	1300 ± 170
Ethyl octanoate	1440	MS, PI	35 ± 2	98 ± 8	11 ± 4	287 ± 35	81 ± 17	249 ± 30
Sum of areas			*715*	*1538*	*391*	*2867*	*1131*	*2049*
Total sum of areas			*11450*	*13935*	*15757*	*27035*	*25393*	*34952*

Data are expressed as the mean ± standard deviation. [a] Linear Retention Index in cheese samples using a StabilWax Column. [b] Identification method: MS = identification by spectra comparison in NIST Library; PI = comparison with published LRI; T = tentatively identified; ST = standard injection.

Among ketones, 2-heptanone was the most abundant ketone identified in our samples at all ripening times, with even higher amounts in longer ripened ones. In particular, 2-heptanone, together with other compounds, was identified among the most important compounds responsible for the characteristic aroma of PR with its fruity and moldy flavor contribution [33]. Ketones can be further transformed in secondary alcohols, and we identified 2-pentanol and 2-heptanol at all ripening times. Even if secondary alcohols are thought to poorly contribute to cheese aroma, 2-heptanol was identified as a key odorant in Grana Padano cheese [39]. Overall, the prolongation of the ripening up to 50 months favored the enrichment of volatile compounds that give the typical aroma to PR cheese, without inducing changes that could generate off-flavors.

3.4. Cheese Microstructure, Appearance and Texture

Regardless the cheese ripening stage, fat was confined in irregular globular-shaped areas trapped within the protein network, suggesting these to originate from coalescence of fat globules of different size (Figure 3). Coalescence evolved during ripening, as confirmed by the lower number of fat globules in most aged cheeses (Table 4), and some small intact globules were detected in the 12mo cheeses only (Figure 3a, arrows). Other parameters, such as sphericity and fat volume, were not significantly affected by the ripening time ($p > 0.05$, data not shown). Similarly, these parameters were not age-associated in Cheddar cheese, since fat distribution established in cheese at manufacturing remained unchanged across ripening [40]. Image analysis showed cheese porosity to increase significantly ($p < 0.05$, one-way ANOVA with respect to time) across ripening from 12 to 50 months. This finding is in agreement with data presented for Cheddar cheese where porosity was considered an effect of proteolysis progression after 30 weeks of ripening [41]. The CLSM images showed the presence of large calcium phosphate crystals in all samples (Figure 3, arrowheads), [11].

Table 4. Image analysis parameters (number of fat globules and porosity), hardness (N) and brittleness (mm) of Parmigiano Reggiano cheeses ripened from 12 to 50 months.

Ripening Time (Months)	Number of Fat Globules	Porosity	Hardness	Brittleness
12	676 ± 28 [c]	0.11 ± 0.01 [a]	11.96 ± 1.57 [a]	5.02 ± 0.34 [a]
18	574 ± 149 [bc]	0.11 ± 0.01 [a]	15.98 ± 0.29 [abc]	6.31 ± 0.38 [ab]
24	292 ± 96 [a]	0.17 ± 0.01 [b]	18.93 ± 0.49 [abcd]	6.52 ± 0.82 [b]
30	410 ± 114 [abc]	0.16 ± 0.01 [b]	21.48 ± 0.49 [bcd]	6.30 ± 0.94 [ab]
40	476 ± 65 [abc]	0.17 ± 0.01 [b]	22.65 ± 4.31 [cbcd]	6.52 ± 0.69 [ab]
50	363 ± 150 [ab]	0.19 ± 0.01 [c]	25.20 ± 3.53 [d]	5.53 ± 0.69 [a]

The results are expressed as the mean ± standard deviation. [a,b,c,d] Mean values within a column with different superscripts are significantly different ($p < 0.05$; Tukey's test). Number of fat globules and porosity were calculated in a volume sample of 212 μm × 212 μm × 7 μm.

Similarly, numerous tyrosine crystals visible with the naked eye were present in 12mo cheeses (Table S2), in agreement with available literature [9,42]. In contrast, the white spots were absent in cheeses at this stage of ripening and appeared in increasing numbers subsequently. White spots are a typical feature of long-ripened hard cheeses, forming as a consequence of slow water migration within the cheese matrix that promotes compartmentalization of the most hydrophobic FAA in restricted areas [11]. This observation suggests that physico-chemical properties of cheese also evolved across an extremely long period.

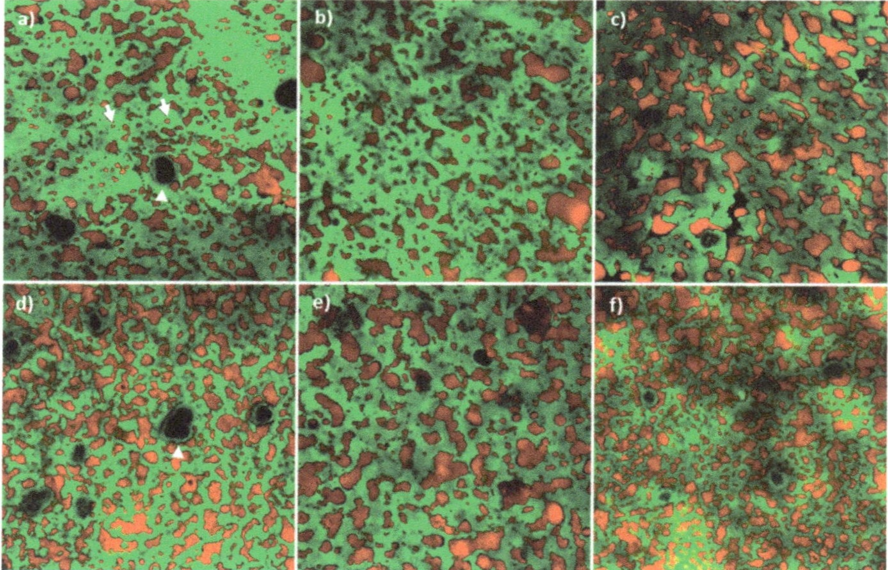

Figure 3. Confocal microscopy images of 12 months (**a**), 18 months (**b**), 24 months (**c**), 30 months (**d**), 40 months (**e**) and 50 months (**f**) ripened Parmigiano Reggiano cheeses. Intact fat globules (white arrows, panel a) were mostly observed in 12-month ripened Parmigiano Reggiano cheese. Calcium phosphate crystals (white arrowheads, panels **a** and **d**) were present at all ripening times.

Curves obtained from texture analysis showed the typical profile of extra-hard cheeses (data not shown). Changes in texture profile were monitored through measurement of both hardness and brittleness. The cheese hardness significantly ($p < 0.05$, one-way ANOVA with respect to time) increased with ripening time (Table 4), differences being significant after 30-month ripening with respect to 12 months. The brittleness instead, did not show a specific trend. Normally, cheese hardness decreases during early ripening as a consequence of proteolysis and hydration of the casein strands, both of which reduce the strength of the casein network. However, loss of moisture in turn causes an increase in protein concentration having an opposite effect [43]. Similar to our results, Noël et al. [9] observed firmness but not brittleness to increase with age in PR cheese during ripening from 12 to 28 months.

3.5. Cheese Color

Color analysis evidenced that the color of PR changed during the ripening period considered (Table 5). The magnitude of color difference between the color indexes recorded at each ripening time and that recorded at 12 months was estimated with ΔE^* values which were all higher than 3, i.e., the threshold for a clearly perceivable color difference [44]. After 30 months of ripening, the difference increased due to the reduction of luminosity (L*) values that passed from an average value of 76 to 70 after 40 and 50 months. This trend could be due to the complex modifications taking place in PR cheese during ripening, especially the moisture reduction as shown in Table 1.

Yellowness index was not related to cheese age. This index is usually associated with seasonal variation of β-carotene and carotenoids content in the diet of dairy cattle [45], and we cannot exclude an initial difference in those components of our samples. Hue angle (h^*) is another color index that can give an indication about the yellow character of the sample; in particular, an angle of 90° represents the yellow hue. While moving towards lower values, the color turns towards pale yellow up to green hue. Hue angle values at 12 and 50 months ripening were very close to 90° and statistically different from the other samples which correspond to lighter yellow hue, in accordance with YI values.

Table 5. Color parameters of Parmigiano Reggiano cheeses ripened from 12 to 50 months.

Ripening Time (Months)	ΔE^* (6)	YI (5)	h^* (7)
12		35.07 [a] ± 0.62	88.65 [a] ± 0.19
18	3.66 [a] ± 0.74	31.42 [a] ± 1.32	82.75 [bc] ± 0.31
24	3.26 [a] ± 1.22	31.04 [a] ± 2.97	82.69 [bc] ± 1.62
30	6.72 [b] ± 1.31	26.28 [b] ± 2.94	80.57 [c] ± 1.07
40	7.73 [b] ± 0.91	26.24 [b] ± 3.02	84.80 [b] ± 0.55
50	6.38 [b] ± 1.50	31.40 [a] ± 2.14	87.64 [a] ± 2.85

The results are expressed as the mean ± standard deviation. [(5)(6)(7)] Equations presented in Section 2.5. [a,b,c] Mean values within a column with different superscripts are significantly different ($p < 0.05$; Tukey's test). ΔE^* = total color difference; YI = yellowness index; h^* = hue angle.

3.6. Principal Component Analysis

Principal component analysis (PCA) was performed to evaluate whether cheese samples were characterized by their ripening periods. A graphic display of loadings and scores is shown in Figure 4. Principal components 1 (PC1) and 2 (PC2) explained 76% and 14% of the total variance, respectively. Cheeses distributed along the PC1, with longer ripened samples (30–50 months) well distinguished from younger ones (12–24 months). Based on the loadings plot, two classes of variables grouped with a strong positive (moisture, αs1-I/ αs1, MNFS and number of fat globules) or a strong negative ($\Sigma\gamma/\Sigma\beta$, butanoic acid, hexanoic acid, heptanoic acid, total VOCs, porosity, hardness and FAA) correlation with the ripening time. These two groups of variables were negatively correlated to each other. Detailed correlation coefficients between the variables are presented in Table S3. Butanoic, hexanoic and heptanoic acids showed a strong positive correlation (r = 0.988, $p < 0.001$; r = 1.000, $p < 0.001$; r = 0.882, $p < 0.05$, respectively) with the total sum of VOCs areas, while the contents of fat, protein and octanoic acid did not show a significant correlation with any other variables ($p > 0.05$).

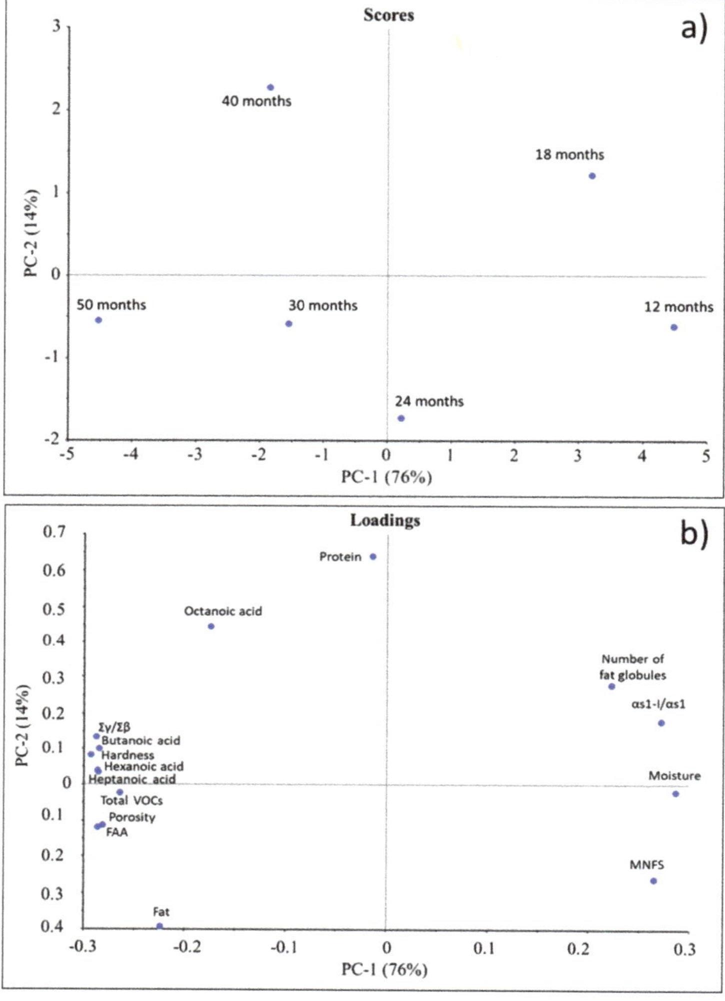

Figure 4. Principal component analysis (**a**), and loadings (**b**) of selected parameters evaluated in Parmigiano Reggiano cheeses ripened up to 50 months.

4. Conclusions

Nowadays, the ripening time is being increased, especially for some hard and extra-hard cheese varieties, to produce premium quality products destined for target markets. Considering that increasing the ripening duration implies parallel increasing costs and, therefore, that these cheeses have a higher price, it was important to elucidate whether changes in cheese characteristics were actually taking place during the extra ripening period. Moisture was the component that changed the most during ripening, even between 40 and 50 months, and a strong negative correlation with cheese hardness was evidenced. In spite of the low water content, both primary and secondary proteolysis proceeded up to 50 months of ripening suggesting a pool of different enzymes still to be active. In parallel, the total volatile fraction increased with a strong contribution of short-chain FFA. Overall, this study showed for the first time that numerous biochemical and structural variations are still ongoing in a hard cheese up to 50 months, proving that the length of ripening deserves to be highlighted to consumers so that they can consciously buy a product with peculiar characteristics that support its premium price.

Supplementary Materials: The following are available online at http://www.mdpi.com/2304-8158/9/3/268/s1, Figure S1: Parmigiano Reggiano manufacture process; Table S1: Free amino acid composition of Parmigiano Reggiano cheeses ripened up to 50 months; Table S2: Number of tyrosine crystals and spots counted in Parmigiano Reggiano cheeses ripened up to 50 months; Table S3: Correlation coefficient matrix.

Author Contributions: P.D., L.P. and S.L. conceived and designed the experiments; P.D., V.R., J.H. and S.G. performed the experiments, collated and analysed the data; P.D. prepared the manuscript; P.D., L.P. and S.L. reviewed the manuscript. All authors have read and agreed to the published version of the manuscript.

Funding: This research received no external funding.

Acknowledgments: The article processing charge was partially covered by the University of Milan. Microscopy observations were carried out at The Advanced Microscopy Facility Platform—UNItech NOLIMITS—University of Milan.

Conflicts of Interest: The authors declare no conflict of interest.

References

1. Consorzio del Formaggio Parmigiano Reggiano. Specification and Regulations. Available online: https://www.parmigianoreggiano.com/consortium/rules_regulation_2/default.aspx (accessed on 5 December 2019).

2. Gatti, M.; Bottari, B.; Lazzi, C.; Neviani, E.; Mucchetti, G. Invited review: Microbial evolution in raw-milk; long-ripened cheeses produced using undefined natural whey starters. *J. Dairy Sci.* **2014**, *97*, 573–591. [CrossRef] [PubMed]

3. Montel, M.-C.; Buchin, S.; Mallet, A.; Delbes-Paus, C.; Vuitton, D.A.; Desmasures, N.; Berthier, F. Traditional cheeses: Rich and diverse microbiota with associated benefits. *Int. J. Food Microbiol.* **2014**, *177*, 136–154. [CrossRef] [PubMed]

4. Sforza, S.; Cavatorta, V.; Lambertini, F.; Galaverna, G.; Dossena, A.; Marchelli, R. Cheese peptidomics: A detailed study on the evolution of the oligopeptide fraction in Parmigiano-Reggiano cheese from curd to 24 months of aging. *J. Dairy Sci.* **2012**, *95*, 3514–3526. [CrossRef] [PubMed]

5. Masotti, F.; Hogenboom, J.A.; Rosi, V.; De Noni, I.; Pellegrino, L. Proteolysis indices related to cheese ripening and typicalness in PDO Grana Padano cheese. *Int. Dairy J.* **2010**, *20*, 352–359. [CrossRef]

6. Malacarne, M.; Summer, A.; Franceschi, P.; Formaggioni, P.; Pecorari, M.; Panari, G.; Mariani, P. Free fatty acid profile of Parmigiano-Reggiano cheese throughout ripening: Comparison between the inner and outer regions of the wheel. *Int. Dairy J.* **2009**, *19*, 637–641. [CrossRef]

7. Cattaneo, S.; Hogenboom, J.A.; Masotti, F.; Rosi, V.; Pellegrino, L.; Resmini, P. Grated Grana Padano cheese: New hints on how to control quality and recognize imitations. *Dairy Sci. Technol.* **2008**, *88*, 595–605. [CrossRef]

8. De Dea Lindner, J.; Bernini, V.; De Lorentiis, A.; Pecorari, A.; Neviani, E.; Gatti, M. Parmigiano–Reggiano cheese: Evolution of cultivable and total lactic microflora and peptidase activities during manufacture and ripening. *Dairy Sci. Technol.* **2008**, *88*, 511–523. [CrossRef]

9. Noël, Y.; Zannoni, M.; Hunter, E.A. Texture of Parmigiano Reggiano cheese: Statistical relationships between rheological and sensory variates. *Le Lait* **1996**, *76*, 243–254. [CrossRef]

10. Zannoni, M. Evolution of the sensory characteristics of Parmigiano-Reggiano cheese to the present day. *Food Qual. Prefer.* **2010**, *21*, 901–905. [CrossRef]

11. D'Incecco, P.; Limbo, S.; Faoro, F.; Hogenboom, J.; Rosi, V.; Morandi, S.; Pellegrino, L. New insight on crystal and spot development in hard and extra-hard cheeses: Association of spots with incomplete aggregation of curd granules. *J. Dairy Sci.* **2016**, *99*, 6144–6156. [CrossRef]

12. D'Incecco, P.; Brasca, M.; Rosi, V.; Morandi, S.; Ferranti, P.; Picariello, G.; Pellegrino, L. Bacterial proteolysis of casein leading to UHT milk gelation: An applicative study. *Food Chem.* **2019**, *292*, 217–226. [CrossRef] [PubMed]

13. Hogenboom, J.A.; D'Incecco, P.; Fuselli, F.; Pellegrino, L. Ion-Exchange Chromatographic Method for the Determination of the Free Amino Acid Composition of Cheese and Other Dairy Products: An Inter-Laboratory Validation Study. *Food Anal. Methods* **2017**, *10*, 3137–3148. [CrossRef]

14. Pathare, P.B.; Opara, U.L.; Al-Said, F.A. Color measurement and analysis in fresh and processed foods: A review. *Food Bioprocess Technol.* **2013**, *6*, 36–60. [CrossRef]

15. Romani, S.; Sacchetti, G.; Pittia, P.; Pinnavaia, G.G.; Dalla Rosa, M. Physical, chemical, textural and sensorial changes of portioned Parmigiano Reggiano cheese packed under different conditions. *Food Sci. Technol. Int.* **2002**, *8*, 203–211. [CrossRef]

16. Ong, L.; D'Incecco, P.; Pellegrino, L.; Nguyen, H.T.; Kentish, S.E.; Gras, S.L. The Effect of Salt on the Structure of Individual Fat Globules and the Microstructure of Dry Salted Cheddar Cheese. *Food Biophys.* **2019**. [CrossRef]

17. D'Incecco, P.; Ong, L.; Gras, S.; Pellegrino, L. A fluorescence in situ staining method for investigating spores and vegetative cells of Clostridia by confocal laser scanning microscopy and structured illuminated microscopy. *Micron* **2018**, *110*, 1–9. [CrossRef]

18. Malacarne, M.; Summer, A.; Panari, G.; Pecorari, M.; Mariani, P. Physico-chemical characterization of Parmigiano-Reggiano cheese ripening (in Italian; abstract in English). *Sci. Tec. Latt. Casearia* **2006**, *57*, 215–228.

19. Fox, P.F.; Guinee, T.P.; Cogan, T.M.; McSweeney, P.L. Microbiology of cheese ripening. In *Fundamentals of Cheese Science*; Springer: Boston, MA, USA, 2017; pp. 333–390.

20. Fenelon, M.A.; Guinee, T.P. Primary proteolysis and textural changes during ripening in Cheddar cheeses manufactured to different fat contents. *Int. Dairy J.* **2000**, *10*, 151–158. [CrossRef]

21. Lopez, C.; Briard-Bion, V.; Beaucher, E.; Ollivon, M. Multiscale characterization of the organization of triglycerides and phospholipids in Emmental cheese: From the microscopic to the molecular level. *J. Agric. Food Chem.* **2008**, *56*, 2406–2414. [CrossRef]

22. Codex Alimentarius. Milk and Milk Products. Available online: http://www.fao.org/3/i2085e/i2085e00.pdf (accessed on 5 December 2019).

23. McSweeney, P.L.H. Biochemistry of cheese ripening. *Int. J. Dairy Technol.* **2004**, *57*, 127–144. [CrossRef]

24. Vélez, M.A.; Bergamini, C.V.; Ramonda, M.B.; Candioti, M.C.; Hynes, E.R.; Perotti, M.C. Influence of cheese making technologies on plasmin and coagulant associated proteolysis. *LWT Food Sci. Technol.* **2015**, *64*, 282–288. [CrossRef]

25. Hynes, E.R.; Aparo, L.; Candioti, M.C. Influence of residual milk-clotting enzyme on αs_1-casein hydrolysis during ripening of Reggianito Argentino cheese. *J. Dairy Sci.* **2004**, *87*, 565–573. [CrossRef]

26. Mayer, H.K.; Rockenbauer, C.; Mlcak, H. Evaluation of proteolysis in Parmesan cheese using electrophoresis and HPLC. *Le Lait* **1998**, *78*, 425–438. [CrossRef]

27. Pellegrino, L.; Rosi, V.; D'Incecco, P.; Stroppa, A.; Hogenboom, J.A. Changes in the soluble nitrogen fraction of milk throughout PDO Grana Padano cheese-making. *Int. Dairy J.* **2015**, *47*, 128–135. [CrossRef]

28. Brasca, M.; Hogenboom, J.A.; Morandi, S.; Rosi, V.; D'Incecco, P.; Silvetti, T.; Pellegrino, L. Proteolytic activity and production of γ-aminobutyric acid by *Streptococcus thermophilus* cultivated in microfiltered pasteurized milk. *J. Agric. Food Chem.* **2016**, *64*, 8604–8614. [CrossRef]

29. D'Incecco, P.; Gatti, M.; Hogenboom, J.A.; Bottari, B.; Rosi, V.; Neviani, E.; Pellegrino, L. Lysozyme affects the microbial catabolism of free arginine in raw-milk hard cheeses. *Food Microbiol.* **2016**, *57*, 16–22. [CrossRef]

30. Pedersen, T.B.; Vogensen, F.K.; Ardö, Y. Effect of heterofermentative lactic acid bacteria of DL-starters in initial ripening of semi-hard cheese. *Int. Dairy J.* **2016**, *57*, 72–79. [CrossRef]

31. Khattab, A.R.; Guirguis, H.A.; Tawfik, S.M.; Farag, M.A. Cheese ripening: A review on modern technologies towards flavor enhancement; process acceleration and improved quality assessment. *Trends Food Sci. Technol.* **2019**, *88*, 343–360. [CrossRef]

32. Frank, D.C.; Owen, C.M.; Patterson, J. Solid phase microextraction (SPME) combined with gas-chromatography and olfactometry-mass spectrometry for characterization of cheese aroma compounds. *LWT Food Sci. Technol.* **2004**, *37*, 139–154. [CrossRef]

33. Qian, M.C.; Burbank, H.M. Hard Italian Cheeses: Parmigiano-Reggiano and Grana Padano. In *Improving the Flavor of Cheese*; Weimer, B.C., Ed.; Woodhead Publishing: Cambridge, UK, 2007; pp. 421–443.

34. Kilcawley, K.N. Cheese flavor. In *Fundamentals of Cheese Science*; Fox, P.F., Guinee, T.P., Cogan, T.M., McSweeney, P.L., Eds.; Springer: Boston, MA, USA, 2017; pp. 443–474.

35. Moio, L.; Addeo, F. Grana Padano cheese aroma. *J. Dairy Res.* **1998**, *65*, 317–333. [CrossRef]

36. Abbatangelo, M.; Núñez-Carmona, E.; Sberveglieri, V. Application of a novel S3 nanowire gas sensor device in parallel with GC-MS for the identification of Parmigiano Reggiano from US and European competitors. *J. Food Eng.* **2018**, *236*, 36–43. [CrossRef]

37. Lazzi, C.; Povolo, M.; Locci, F.; Bernini, V.; Neviani, E.; Gatti, M. Can the development and autolysis of lactic acid bacteria influence the cheese volatile fraction? The case of Grana Padano. *Int. J. Food Microbiol.* **2016**, *233*, 20–28. [CrossRef] [PubMed]

38. Ceruti, R.J.; Zorrilla, S.E.; Sihufe, G.A. Volatile profile evolution of Reggianito cheese during ripening under different temperature–time combinations. *Eur. Food Res. Technol.* **2016**, *242*, 1369–1378. [CrossRef]

39. Curioni, P.M.G.; Bosset, J.O. Key odorants in various cheese types as determined by gas chromatography-olfactometry. *Int. Dairy J.* **2002**, *12*, 959–984. [CrossRef]

40. Soodam, K.; Ong, L.; Powell, I.B.; Kentish, S.E.; Gras, S.L. Effect of elevated temperature on the microstructure of full fat Cheddar cheese during ripening. *Food Struct.* **2017**, *14*, 8–16. [CrossRef]

41. Soodam, K.; Ong, L.; Powell, I.B.; Kentish, S.E.; Gras, S.L. Effect of calcium chloride addition and draining pH on the microstructure and texture of full fat Cheddar cheese during ripening. *Food Chem.* **2015**, *181*, 111–118. [CrossRef]

42. Tansman, G.F.; Kindstedt, P.S.; Hughes, J.M. Crystal fingerprinting: Elucidating the crystals of Cheddar, Parmigiano-Reggiano, Gouda, and soft washed-rind cheeses using powder x-ray diffractometry. *Dairy Sci. Technol.* **2015**, *95*, 651–664. [CrossRef]

43. Fox, P.F.; Guinee, T.P.; Cogan, T.M.; McSweeney, P.L. Cheese: Structure; Rheology and Texture. In *Fundamentals of Cheese Science*; Springer: Boston, MA, USA, 2017; pp. 475–532.

44. Mokrzycki, W.S.; Tatol, M. Color difference Delta E—A survey. *Mach. Graph. Vis.* **2011**, *20*, 383–411.

45. Cerquaglia, O.; Sottocorno, M.; Pellegrino, L.; Ingi, M. Detection of cow's milk; fat or whey in ewe and buffalo ricotta by HPLC determination of β-carotene. *Ital. J. Food Sci.* **2011**, *23*, 367.

© 2020 by the authors. Licensee MDPI, Basel, Switzerland. This article is an open access article distributed under the terms and conditions of the Creative Commons Attribution (CC BY) license (http://creativecommons.org/licenses/by/4.0/).

Article

Quantification of Cheese Yield Reduction in Manufacturing Parmigiano Reggiano from Milk with Non-Compliant Somatic Cells Count

Piero Franceschi [1], Michele Faccia [2], Massimo Malacarne [1,*], Paolo Formaggioni [1,*] and Andrea Summer [1]

1   Department of Veterinary Science, University of Parma, Via del Taglio 10, I-43126 Parma, Italy; piero.franceschi@unipr.it (P.F.); andrea.summer@unipr.it (A.S.)
2   Department of Soil, Plant and Food Sciences, University of Bari. Via Amendola 165/A, 70125 Bari, Italy; michele.faccia@uniba.it
*   Correspondence: massimo.malacarne@unipr.it (M.M.); paolo.formaggioni@unipr.it (P.F.); Tel.: +39-0521032615 (M.M.); +39-0521032614 (P.F.)

Received: 15 January 2020; Accepted: 15 February 2020; Published: 18 February 2020

Abstract: The mammary gland inflammation process is responsible for an increased number of somatic cells in milk, and transfers into the milk of some blood components; this causes alterations in the chemical composition and physico-chemical properties of milk. For this reason, somatic cell count (SCC) is one of the most important parameters of milk quality; therefore, European Union (EU) Regulation no 853/2004 has stated that it must not exceed the limit value of 400,000 cells/mL. The research aimed to compare chemical composition, cheese yield, and cheesemaking losses of two groups of vat milks used for Parmigiano Reggiano production, characterized by different SCC levels. During two years, ten cheesemaking trials were performed in ten different cheese factories. In each trial, two cheesemaking processes were conducted in parallel: one with low SCC milk (below 400,000 cells/mL; Low Cell Count (LCC)) and the other with high SCC milk (400,000–1,000,000 cells/mL; High Cell Count (HCC)). For each trial, vat milk and cooked whey were analyzed; after 24 months of ripening, cheeses were weighed to calculate cheese yield. The HCC group had lower casein content (2.43 vs. 2.57 g/100 g; $p \leq 0.05$) and number (77.03% vs. 77.80%; $p \leq 0.05$), lower phosphorus (88.37 vs. 92.46 mg/100g; $p \leq 0.05$) and titratable acidity (3.16 vs. 3.34 °SH/50 mL; $p \leq 0.05$) compared to LCC. However, chloride (111.88 vs. 104.12 mg/100 g; $p \leq 0.05$) and pH (6.77 vs. 6.71; $p \leq 0.05$) were higher. Fat losses during cheesemaking were higher (20.16 vs. 16.13%). After 24 months of ripening, cheese yield was 8.79% lower for HCC milk than LCC (6.74 vs. 7.39 kg/100 kg; $p \leq 0.05$).

Keywords: Parmigiano Reggiano cheese; somatic cells; milk composition; cheese yield; cheesemaking losses

1. Introduction

Mastitis is the inflammation of the mammary gland caused by bacterial infection. As a response to the inflammation, the number of macrophages, leucocytes, and polymorphonuclear cells strongly increases, causing a high level of the somatic cells in milk [1]. Somatic cell count (SCC) is one of the most important parameters of milk quality, both under the safety and technological point of views. As to safety, a high SCC level indicates poor hygienic quality and possible presence of pathogens. In order to protect the consumer's health, the law (European Union (EU) Regulation no 853/2004) [2] has regulated this parameter for cow milk. In particular, the value (expressed as rolling geometric mean calculated over a period of three months, with at least one sampling per month) must not exceed the limit value of 400,000 cells/mL. Besides hygienic concerns, a high SCC level negatively influences the technological properties of milk, and both the coagulation process and the chemical-sensory characteristics of the cheeses tend to worsen. Several authors have investigated the causes of technological worsening, and they can be summarized in increased levels of casein passing into the soluble phase, greater proteolytic activity, and modifications of the balance of mineral salts [3–5].

An important issue connected with the application of Regulation 853/2004 [2] is that the limit of 400,000 cells/mL is calculated over a three-month period. During this period, the risk of milk not complying the legal limit that reaches the dairy transformation, does exist. Summer et al. [6] reported that 10.71% of the samples collected from free stalls in the Parmigiano Reggiano production area (Italy), exceeded the legal limit of milk production from 2006 to 2008. This was confirmed by data of the Lombardia and Emilia Romagna Experimental Zootechnical Institute (IZSLER, Italy). According to these results, about 10% of milk production in 2018 in these areas exceeded the 400,000 cells/mL limit [7]. These milks are mainly concentrated in the summer period [6], when the cows are subject to stress due to the hot and humid climate typical of the Po valley plain [8].

Two of the most produced cheeses in Lombardy and Emilia-Romagna regions are Grana Padano and Parmigiano Reggiano. Both have been recognized as EU Protected Designation of Origin (PDO) products [9] and are very similar from a chemical, technological, and nutritional point of view [10], being hard cooked cheeses manufactured by partially skimmed raw milk, added with autochthonous starter culture. According to the official protocol, partially skimmed milk from evening milking is merged with the full cream morning milk (about 50:50 *v/v*), giving rise to "vat milk". Previous studies conducted on Parmigiano Reggiano reported wide variations of the composition of the vat milk throughout the year, and a significant decrease of the cheese yield when the milk somatic cells count exceeded 300,000 cells/mL [11,12]. Considering the importance of cheese yield on the economic efficiency of Parmigiano Reggiano cheesemaking, a specific study aiming to quantify the cheese yield capacity of milk with different levels of somatic cells is highly required. Such information could give a useful contribution to quantify the right remuneration of milk, motivating the breeders to make investments for reducing the somatic cells content. The aim of the present research is to compare chemical composition, cheese yield, and cheesemaking losses (during manufacturing to Parmigiano Reggiano) of milk with non-law-complying and law-complying SCC.

2. Materials and Methods

2.1. Parmigiano Reggiano Cheesemaking Process

Cheeses were produced according to the official protocol of Parmigiano Reggiano PDO cheese. Although the protocol does not include breed restrictions and allows commingling milk coming from different farms, the milk used in the experimentations was collected only from Italian Friesian cattle herds. Moreover, each cheesemaking trial was performed using milk derived from a single herd.

Vat milk was obtained by commingling the partially skimmed evening milk and the full cream morning milk. In brief, the whole milk of evening milking (WE-milk) was collected from the farm and transported to the cheese factory where it was placed into the creaming tank. Overnight, the cream was separated naturally from the milk and was collected. The morning after, the partially skimmed milk was obtained (fat content about 1.5 g 100 g^{-1}), extracted from the bottom of the tank, and transferred into the cheesemaking vat; the same amount of the whole milk that was obtained by the morning milking of the same herd was added. The final merged milk was lower in fat, containing about 2.6 g 100 g^{-1} fat, with a fat to casein ratio of about 1–1.1. It is called vat milk (V-milk) and, before undergoing the cheesemaking process, is added with natural whey starter culture (2.5–3 liters 100 kg^{-1}). The whey starter culture was obtained by spontaneous acidification of cooked whey (C-whey), deriving from the cheesemaking of the previous day. The inoculated vat milk was heated to 33 °C and clotted with 2.5–3 g 100 kg^{-1} of commercial calf rennet (1:120,000 strength). After 10–12 min, coagulation occurred; the curd was then cut into small granules (having approximately the size of a rice grain), and heated by increasing temperature to 55 °C. After cooking, the small curd particles were left to deposit at the bottom of the vat by decantation, and a whole curd mass was formed. During this time, the temperature remained approximately 53–55 °C. The curd mass was then removed from the vat, divided into two parts, and placed into molds for two day. They were periodically turned over to promote syneresis and whey draining. The cheese wheels were then salted in brine for a period of 20 days before entering the ripening room, where they remained for at least 24 months.

2.2. Experimental Design

Over a period of two years, ten comparative cheesemaking trials were performed in ten different cheese factories located in the Parmigiano Reggiano production area. In each trial, two cheesemaking processes were performed in parallel: one using V-milk coded as Low Cell Count (LCC) and the other using V-milk coded as High Cell Count (HCC). The classification LCC or HCC is based on SCC of the whole evening milk: LCC contained less than 400,000 cells/mL, whereas HCC contained more than 400,000 cells/mL, but less than 1,000,000 cells/mL.

2.3. Analytical Methods

In each cheesemaking trial, the following type of samples were collected: WE-milk taken from the tank of the farm; V-milk, taken from the vat before the addition of the natural whey starter; C-whey, taken from the vat after the extraction of the cheese, and stirred for 5 min. Sodium merthiolate (0.02 g/100 mL, as preservative) was added to all samples, which were cooled to 5 °C, and immediately transported to the laboratory for analyses.

The following tests were performed: somatic cells on WE- and V-milk using the fluoro-opto-electronic method with Fossomatic (Foss Electric, Hillerød, Denmark) [13]; fat content on V-milk by mid-infrared spectroscopy [14] with Milko-Scan FT6000 (Foss Electric); fat content of C-whey was determined by the volumetric Gerber method [15].

Total N (TN), non-casein N (NCN), and non-protein N (NPN) on V-milk, acid whey at pH 4.6 and Trichloroacetic acid (TCA 120 g/L; Carlo Erba Reagents, Milan, Italy) filtered whey, respectively, were determined by the Kjeldahl method, which was performed using a DK6 Digestor and UDK126A Distiller (VELP Scientifica, Usmate, Italy), according to the Association of Official Analytical Chemists (AOAC) standards [16–18]. From these analyses, crude protein (TN × 6.38/1000), casein ((TN-NCN) × 6.38/1000), casein number ((TN-NCN) × 100/TN), NPN × 6.38 (NPN × 6.38/1000), true protein ((TN-NPN) × 6.38/100) were calculated, as described by Summer et al. [12]. TN was also determined on C-whey, also by Kjeldahl [12].

The content of phosphorus was measured on V-milk and C-whey by colorimetric method [19] and those of calcium and magnesium by atomic absorption spectrometry (AAS Perkin-Elmer 1100 B, Waltham, MA, USA), as reported by Malacarne et al. [19]; chloride were determined on V-milk by argentometric Volhard method [20].

The pH was measured on V-milk with potentiometer Crison (Crison Instruments, E-08328 Barcelona, Spain), and titratable acidity by titration with sodium hydroxide solution (0.25 N) of 50 mL of V-milk using as indicator 2 mL of phenolphthalein in ethanol (20 g/L; Carlo Erba Reagents), according to the Soxhlet-Henkel method [12].

Somatic cell score (SCS = (Log$_2$(SCC/100) + 3) [21] and fat-to-casein ratio values were calculated on WE-milk and V-milk, respectively.

For yield calculation, in each cheesemaking, V-milk was weighed directly in the vat, before the addition of the starter. Both the two cheese wheels resulting from each cheesemaking were weighed at 24 months of ripening; the moisture content of the cheese after drying at 102 °C was also determined [22].

The actual cheese yield (ACY; kg of cheese /100 kg of milk) was calculated as follows:

$$ACY = cw \times 100/mw,$$

where: ACY = actual cheese yield; cw = cheese weight, expressed in kg; mw = milk weight, also expressed in kg.

The adjusted dry cheese yield (ADY; kg of cheese /100 kg of milk) was calculated, according to the following formula:

$$ADY = ACY \times (100 - CMC)/100,$$

where: ADY = adjusted yield; CMC = cheese moisture content.

Finally, the estimated cheesemaking losses (ECL) values of protein, casein, fat, calcium, phosphorus, and magnesium were calculated as follows:

$$ECL = (C\text{-whey}) \times 100/(V\text{-milk}),$$

where ECL is expressed as percentage; (C-whey) = concentration in whey, expressed as g/100 g (mg/100 g for Ca, P, Mg); (V-milk) = concentration in milk, expressed as g/100 g (mg/100 g for Ca, P, Mg).

2.4. Statistical Analysis

The significance of the differences between classes (LCC and HCC) was tested by analysis of variance, using the general linear model procedure of SPSS (IBM SPSS Statics 24, Armonk, NY, USA), according to the following univariate model:

$$Y_{ijk} = \mu + C_i + T_j + \varepsilon_{ijk},$$

where: Y_{ijk} = dependent variable; μ = overall mean; C_i = effect of milk somatic cell class (LCC or HCC) (i = 1, 2); T_j = effect of trial (j = 1, … .10); ε_{ijk} = residual error. The significance of the differences was tested by least significant differences method.

Data were also processed by the Pearson product moment correlation coefficient, to measure the degree of the linear relationship between the somatic cell content of the whole evening milk (expressed as somatic cell score) and cheese yield, and between the vat milk chemical characteristics and cheese yield.

3. Results and Discussion

3.1. Chemical Composition and Physico-Chemical Properties of Vat Milk

The average somatic cell count in LCC WE-milk was 233,000 cells/mL (*minimum* 122,000; *maximum* 341,000 cells/mL) and 538,000 cells/mL (*minimum* 407,000; *maximum* 886,000 cells/mL) in HCC WE-milk. The fat and protein contents were 3.70 and 3.61 g/100 g, and 3.18 and 3.06 g/100 g, respectively for LCC and HCC-milk.

Chemical composition, physicochemical properties, and somatic cells count of LCC and HCC vat milk are described in Table 1. Casein, casein number, phosphorus, chloride, titratable acidity, and pH showed different values, with $p \leq 0.05$.

The HCC V-milk, in comparison with the LCC V-milk, was characterized by lower casein (2.43 vs. 2.57 g/100 g; $p \leq 0.05$) and phosphorus (88.37 vs. 92.46 mg/100 g; $p \leq 0.05$) contents, lower casein number value (77.03 vs. 77.80%; $p \leq 0.05$) and higher chloride content (111.88 vs. 104.12 mg/100 g; $p \leq 0.05$). Moreover, titratable acidity was lower (3.16 vs. 3.34 °SH/50 mL; $p \leq 0.05$) and the pH value was higher (6.77 vs. 6.71 units; $p \leq 0.05$). Overall, the composition profile was less favorable for cheesemaking. In fact, cheese yield strictly depends on fat and casein contents [23], and pH, acidity, and mineral contents of the milk are closely linked to the rennet-coagulation properties [24].

The results are in accordance with those reported in the literature and collected in a review by Le Maréchal et al. [1]. More specifically, Franceschi et al. [11], in research performed on 248 Parmigiano Reggiano vat milk samples (202 with less than 400,000 cells/mL and 48 with more than 400,000 cells/mL), reported lower casein content (2.43 vs. 2.47 g/100 g $p \leq 0.05$), casein number (76.78 vs. 77.42 $p \leq 0.05$), and titratable acidity (3.24 vs. 3.29 °SH/50 mL $p \leq 0.05$) for milk with more than 400,000 cells/mL than milk with less than 400,000. Summer et al. [25] observed, in a study conducted on 26 single quarter milk samples (13 with less and 13 with more than 400,000 cells/mL), a lower phosphorous content (87.76 vs. 89.71 mg/100 g $p \leq 0.05$) and a higher chloride content (135.17 vs. 99.43 mg/100 g $p \leq 0.01$).

All these changes in HCC milk are due to an increase of the concentration of whey protein [26,27] and sodium chloride [25], which come directly from blood [27], and a decrease of phosphorus [12,25] and casein [10], caused by decreased mammary activity and an increase of proteolytic enzymatic activity [26,28]. Being that titratable acidity and pH were highly correlated with the contents of casein and phosphorus, the significant lower and higher values of these two parameters, respectively, in HCC V-milk are explained.

Table 1. Chemical composition and physicochemical properties parameters of vat milk with less than 400,000 (Low Cell Count (LCC)) and more than 400,000 cells/mL (High Cell Count (HCC)) (least square mean values ± standard error).

Parameter	Unit of Measure	LCC [1] n [2] $= 10$	HCC [1] n [2] $= 10$	SE [3]	p [4]
Crude protein	g/100 g	3.30	3.16	0.06	NS
Casein	g/100 g	2.57	2.43	0.05	*
Casein number	%	77.80	77.03	0.31	*
NPN × 6.38	g/100 g	0.17	0.16	0.01	NS
True protein	g/100 g	3.13	3.00	0.06	NS
Fat	g/100 g	2.75	2.68	0.05	NS
Fat to casein ratio	Value	1.07	1.10	0.01	NS
Calcium	mg/100 g	114.75	114.17	1.54	NS
Phosphorus	mg/100 g	92.46	88.37	1.32	*
Magnesium	mg/100 g	11.12	10.70	0.25	NS
Chloride	mg/100 g	104.12	111.88	2.72	*
Titratable acidity	°SH/50 mL	3.34	3.16	0.04	*
pH	Value	6.71	6.77	0.02	*
Somatic cell count	10^3 cells/mL	146	259	5	**

[1] Classification was based on somatic cell count (SCC) of the whole evening milk; [2] Number of samples; [3] Standard error; [4] p-value: NS, $p > 0.05$; * $p \leq 0.05$; ** $p \leq 0.01$.

3.2. Cheese Yield and Cheesemaking Losses of Vat Milk

The least square mean values of cheese yield are shown in Table 2. Both actual cheese yields at 24 months of ripening and dry yield at 24 months of ripening were significantly lower in HCC cheesemaking than LCC (Figure 1).

In the literature, there is a general consensus about the negative relationships between somatic cells of milk and its cheese yield capacity. In hard Cheddar cheese production, a decrease of the yield was reported for milk with more than 100,000 cells/mL [4] and more than 300,000 cells/mL [29]. Similarly, in cottage (a soft cheese) production, Klei et al. [5] found that cheese yield efficiency was 4.34% lower for high somatic cell milk (mean value 872,000 cells/mL) than for low somatic cell milk (83,000 cells/mL). In contrast, Mazal et al. [30] observed no significant differences in Prato cheese yield when comparing milk with less than 200,000 cells/mL and more than 600,000 cells/mL (10.4 vs. 9.2 kg/100 kg, respectively). Such different results could depend on the low number of comparison made (only three comparative trials) and differences in cheesemaking technologies between Prato cheese (a soft cheese made with pasteurized milk) and hard cheeses. For Parmigiano Reggiano, the decrease of yield measured 24 h after the extraction of the curd started at 300,000 cells/mL [12]. The difference of the mean value of the 24-month cheese yield between the LCC milk and HCC was 0.65 kg/100 kg of processed milk, corresponding to a difference of 9.64%.

Table 2. Cheese yield and cheesemaking losses of vat milk with less than 400,000 (LCC) and more than 400,000 cells/mL (HCC) (least square mean values ± standard error).

Parameter	Unit of Measure	LCC [1] n [2] = 10	HCC [1] n [2] = 10	SE [3]	p [4]
Cheese yield:					
Actual cheese yield at 24 months	kg/100 kg	7.39	6.74	0.18	*
Dry yield at 24 months	kg/100 kg	5.19	4.74	0.20	*
Cheese characteristics:					
Moisture	g/100 g	29.84	29.78	0.28	NS
Estimated cheesemaking losses:					
Protein	%	26.59	26.92	0.28	NS
Casein	%	5.65	5.13	0.22	NS
Fat	%	16.13	20.16	0.87	**
Phosphorus	%	49.86	50.28	0.87	NS
Calcium	%	33.86	34.71	0.51	NS
Magnesium	%	76.92	77.36	1.69	NS

[1] Classification was based on SCC of the whole evening milk; [2] Number of samples; [3] Standard error; [4] p-value: NS, $p > 0.05$; * $p \leq 0.05$; ** $p \leq 0.01$.

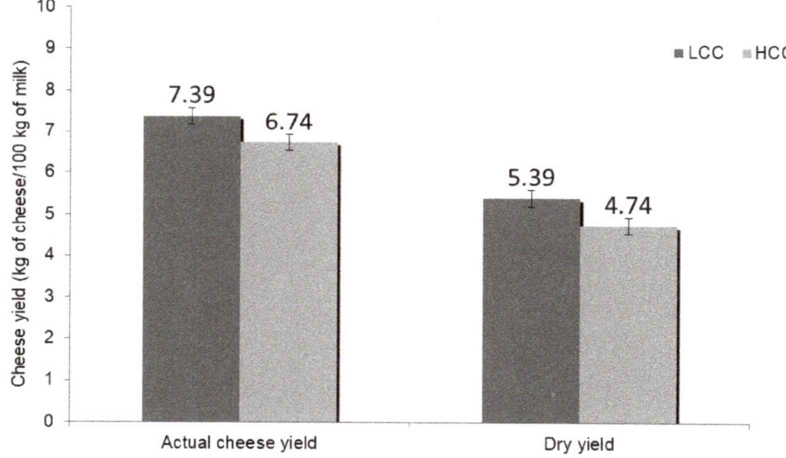

Figure 1. Cheese yield at 24 months ripening from milk with less than 400,000 (Low Cell Count (LCC)) and more than 400,000 cells/mL (High Cell Count (HCC)) (least square mean values). Classification was based on SCC of the whole evening milk. For both, LCC and HCC means differ with a p-value ≤ 0.05.

The difference in cheese yield were not caused by differences in water retention (the moisture of the cheeses was the same); this was confirmed by the calculation of the adjusted dry yield. In effect, cheese yield is directly proportional to casein and fat content [23]; moreover, the worsening of rennet coagulation properties [31], linked to the decrease in phosphorus and calcium, and to the reduction of the casein content itself, leads to an increase in the loss of fat in the whey.

In Table 2, fat, protein, and main minerals estimated cheesemaking losses are also shown. Only for fat losses, different average values ($p \leq 0.01$) were observed between LCC and HCC V-milks, with higher losses in HCC-cheesemaking than in LCC (Figure 2). In a recent paper, Franceschi et al. [31] reported that the average fat loss in Parmigiano Reggiano, calculated on 288 cheesemaking trials, was 16.93%. In the present study, the fat loss in LCC-cheesemaking was in agreement with this value, while fat loss of the HCC-cheesemaking is much higher than in Franceschi et al. [31]. Poor efficiency of HCC curd in retaining fat into the casein reticulum may be both related to undesired structural modification of the curd and/or loss of integrity of the fat globule, resulting from the increased activities of plasmin and lipases in milk collected from infected-glands [4,32]. This observation was confirmed by the positive and significant correlation found between milk casein content and fat losses and between milk fat to casein ratio and fat losses (Table 3).

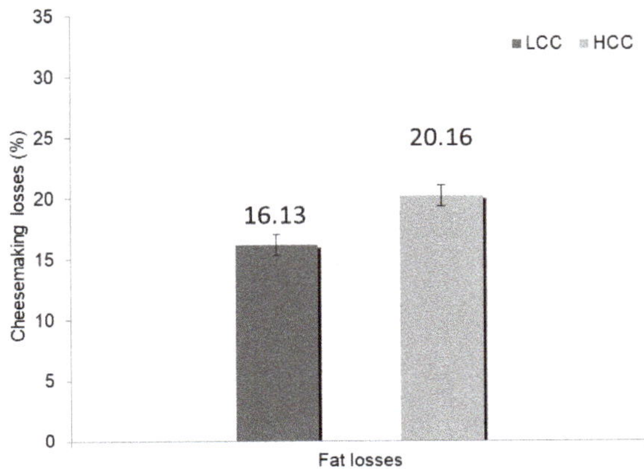

Figure 2. Cheesemaking fat losses of milk with less than 400,000 (LCC) and more than 400,000 cells/mL (HCC) (least square mean values). Classification was based on somatic cell count (SCC) of the whole evening milk. LCC and HCC means differ with a p-value ≤ 0.01.

Table 3. Pearson correlation coefficient (r) between somatic cell content of whole evening milk, cheese yield and cheesemaking losses, and between the vat milk chemical parameters, cheese yield and cheesemaking losses.

	Cheese Yield [1]		Protein Losses		Fat Losses	
	r	p [2]	r	p [2]	r	p [2]
Somatic cells [3]	−0.57	*	0.06	NS	0.34	NS
Crude protein	0.68	*	0.03	NS	−0.58	**
Casein	0.69	*	−0.20	NS	−0.64	**
Fat	0.60	*	−0.09	NS	−0.34	NS
Fat to casein ratio	−0.18	NS	0.15	NS	0.46	*

[1] At 24 months ripening; [2] p-value: NS, $p > 0.05$; * $p \leq 0.05$; ** $p \leq 0.01$; [3] Expressed as somatic cell score (SCS).

Pearson product moment correlation coefficients between WE-milk somatic cell content, cheese yield and cheesemaking losses, and between V-milk characteristics, cheese yield, and cheesemaking losses, are shown in Table 3.

A negative and significant correlation was observed between SCC (measured on the corresponding WE-milk before natural creaming) and V-milk cheese yield at 24 months ripening. Conversely, V-milk cheese yields were positively correlated with contents of protein, casein, and fat. Formaggioni et al. [23] reported, for their prediction formulas of the estimated cheese yield for hard cooked cheese types, a similar correlation between Parmigiano Reggiano cheese yield, and the content of milk crude protein, casein, and fat constituents. Furthermore, the same positive correlations are reported in other cheese typologies, as Grana Padano [33] and Saint-Nectaire cheeses [34].

4. Conclusions

In conclusion, the present study allowed quantifying the negative impact of milk somatic cell content on the cheese yield in manufacturing Parmigiano Reggiano cheese. The decrease of the yield both derives from the less favorable chemical composition of milk and from reduced efficiency of the coagulum to retain the fat fraction. It is very important to know that, even though Regulation (EC) no 853/2004 [2] indirectly allows to submit to cheesemaking milk with more than 400,000 cells/mL, due to the limited number of mandatory analyses, this must be avoided. The economic impact of the connected reduction of the cheese yield is not negligible, and could compromise profitability.

Author Contributions: Conceptualization, A.S., P.F. (Piero Franceschi), and M.F.; methodology, A.S., P.F. (Piero Franceschi) and M.M.; software, P.F. (Piero Franceschi) and P.F. (Paolo Formaggioni); validation, A.S., M.M., P.F. (Paolo Formaggioni) and P.F. (Piero Franceschi); formal analysis, P.F. (Piero Franceschi), P.F. (Paolo Formaggioni); investigation, M.M., and A.S.; resources, A.S. and M.M.; data curation, P.F. (Piero Franceschi), P.F. (Paolo Formaggioni) and M.F.; writing—original draft preparation, P.F. (Piero Franceschi), M.F. and P.F. (Paolo Formaggioni).; writing—review and editing, M.F., M.M., and P.F. (Paolo Formaggioni); visualization, P.F. (Piero Franceschi), A.S., M.M., and M.F.; supervision, A.S. and M.F.; project administration, A.S. and M.M. All authors have read and agreed to the published version of the manuscript.

Funding: This research received no external funding.

Conflicts of Interest: The authors declare that there are no conflicts of interest in this research article.

References

1. Le Maréchal, C.; Thiéry, R.; Vautor, E.; Le Loir, Y. Mastitis impact on technological properties of milk and quality of milk products—A review. *Dairy Sci. Technol.* **2011**, *91*, 247–282. [CrossRef]
2. Regulation (EC) No 853/2004 of the European Parliament and of the Council of 29 April 2004, Laying Down Specific Hygiene Rules for on the Hygiene of Foodstuffs, Web Site. Available online: https://eur-lex.europa.eu/LexUriServ/LexUriServ.do?uri=OJ:L:2004:139:0055:0205:en:PDF (accessed on 12 December 2019).
3. Ali, A.E.; Andrews, A.T.; Cheeseman, G.C. Influence of elevated somatic cell count on casein distribution and cheese-making. *J. Dairy Res.* **1980**, *47*, 393–400. [CrossRef]
4. Barbano, D.M.; Rasmussen, R.R.; Lynch, J.M. Influence of milk somatic cell count and milk age on cheese yield. *J. Dairy Sci.* **1991**, *74*, 369–388. [CrossRef]
5. Klei, L.; Yun, J.; Sapru, A.; Lynch, J.; Barbano, D.M.; Sears, P.; Galton, D. Effects of milk somatic cell count on Cottage cheese yield and quality. *J. Dairy Sci.* **1998**, *81*, 1205–1213. [CrossRef]
6. Summer, A.; Franceschi, P.; Formaggioni, P.; Malacarne, M. Characteristics of raw milk produced by free-stall or tie-stall cattle herds in the Parmigiano-Reggiano cheese production area. *Dairy Sci. Technol.* **2014**, *94*, 581–590. [CrossRef]
7. Lombardy and Emilia Romagna Experimental Zootechnical Institute (IZSLER) Web Site. Available online: https://www.izsler.it/pls/izs_bs/v3_s2ew_consultazione.mostra_pagina?id_pagina=524 (accessed on 12 December 2019).
8. Summer, A.; Lora, I.; Formaggioni, P.; Gottardo, F. Impact of heat stress on milk and meat production. *Anim. Front.* **2019**, *9*, 39–46. [CrossRef]
9. Council Regulation (EC) No 510/2006 of 20 March 2006 on the Protection of Geographical Indications and Designations of Origin for agricultural Products and Foodstuffs, Web Site. Available online: https://eur-lex.europa.eu/legal-content/EN/TXT/PDF/?uri=CELEX:32006R0510&from=en (accessed on 12 December 2019).
10. Summer, A.; Formaggioni, P.; Franceschi, P.; Di Frangia, F.; Righi, F.; Malacarne, M. Cheese as functional food: The example of Parmigiano Reggiano and Grana Padano. *Food Tech. Biotech.* **2017**, *55*, 277–289. [CrossRef]
11. Franceschi, P.; Summer, A.; Sandri, S.; Formaggioni, P.; Malacarne, M.; Mariani, P. Effects of the full cream milk somatic cell content on the characteristics of vat milk in the manufacture of Parmigiano-Reggiano cheese. *Vet. Res. Commun.* **2009**, *33* (Suppl. 1), 281–283. [CrossRef]
12. Summer, A.; Franceschi, P.; Formaggioni, P.; Malacarne, M. Influence of milk somatic cell content on Parmigiano-Reggiano cheese yield. *J. Dairy Res.* **2015**, *82*, 222–227. [CrossRef]
13. IDF Standard. *Milk, Enumeration of Somatic Cells, Part 2: Guidance on the Operation of Fluoro-Opto-Electronic Counters*; 148-2/ISO13366-2; International Dairy Federation Standard: Brussels, Belgium, 2006.
14. IDF Standard. *Milk and Liquid Milk Products, Guidelines for the Application of Mid- Infrared Spectrometry*; 141/ISO9622; International Dairy Federation Standard: Brussels, Belgium, 2013.
15. IDF Standard. *Milk, Determination of Fat Content, Acido-Butyrometric (Gerber Method)*; 238-2/ISO19662-2; International Dairy Federation Standard: Brussels, Belgium, 2018.

16. Association of Official Analytical Chemists [AOAC]. Nitrogen (total) in milk, method no. 991.20. In *Official Methods of Analysis of AOAC International*, 18th ed.; Horowitz, W., Ed.; AOAC International: Gaithersburg, MD, USA, 2005; pp. 10–12.

17. Association of Official Analytical Chemists [AOAC]. Noncasein nitrogen content of milk, method no. 998.05. In *Official Methods of Analysis of AOAC International*, 18th ed.; Horowitz, W., Ed.; AOAC International: Gaithersburg, MD, USA, 2005; pp. 50–51.

18. Association of Official Analytical Chemists [AOAC]. Nonprotein nitrogen in whole milk, method no. 991.21. In *Official Methods of Analysis of AOAC International*, 18th ed.; Horowitz, W., Ed.; AOAC International: Gaithersburg, MD, USA, 2005; pp. 12–13.

19. Malacarne, M.; Criscione, A.; Franceschi, P.; Tumino, S.; Bordonaro, S.; Di Frangia, F.; Marletta, D.; Summer, A. Distribution of Ca, P and Mg and casein micelle mineralisation in donkey milk from the second to ninth month of lactation. *Int. Dairy J.* **2017**, *66*, 1–5. [CrossRef]

20. Malacarne, M.; Criscione, A.; Franceschi, P.; Bordonaro, S.; Formaggioni, P.; Marletta, D.; Summer, A. New insights into chemical and mineral composition of donkey milk throughout nine months of lactation. *Animals* **2019**, *9*, 1161. [CrossRef] [PubMed]

21. Shook, G.E.; Schutz, M.M. Selection on somatic cell score to improve resistance to mastitis in the United States. *J. Dairy Sci.* **1994**, *77*, 648–658. [CrossRef]

22. IDF Standard. *Cheese and Processed Cheese, Determination of the Total Solids Content (Reference Method)*; 4/ISO5534; International Dairy Federation Standard: Brussels, Belgium, 2004.

23. Formaggioni, P.; Summer, A.; Malacarne, M.; Franceschi, P.; Mucchetti, G. Italian and Italian-style hard cooked cheeses: Predictive formulas for Parmigiano-Reggiano 24 h cheese yield. *Int. Dairy J.* **2015**, *51*, 52–58. [CrossRef]

24. Malacarne, M.; Franceschi, P.; Formaggioni, P.; Sandri, S.; Mariani, P.; Summer, A. Influence of micellar calcium and phosphorus on rennet coagulation properties of cows milk. *J. Dairy Res.* **2014**, *81*, 129–136. [CrossRef] [PubMed]

25. Summer, A.; Franceschi, P.; Malacarne, M.; Formaggioni, P.; Tosi, F.; Tedeschi, G.; Mariani, P. Influence of somatic cell count on mineral content and salt equilibria of milk. *Ital. J. Anim. Sci.* **2009**, *8* (Suppl. 2), 435–437. [CrossRef]

26. Urech, E.; Puhan, Z.; Schällibaum, M. Changes in milk protein fraction as affected by subclinical mastitis. *J. Dairy Sci.* **1999**, *82*, 2402–2411. [CrossRef]

27. Somers, J.; O'Brien, B.; Meany, W.; Kelly, A.L. Heterogeneity of proteolytic enzyme activities in milk samples of different somatic cell count. *J. Dairy Res.* **2003**, *70*, 45–50. [CrossRef]

28. Shennan, D.B.; Peaker, M. Transport of milk constituents by the mammary gland. *Physiol. Rev.* **2000**, *80*, 925–951. [CrossRef]

29. Politis, I.; Ng-Kwai-Hang, K.F. Association between somatic cell count of milk and cheese-yielding capacity. *J. Dairy Sci.* **1988**, *71*, 1720–1727. [CrossRef]

30. Mazal, G.; Vianna, P.C.B.; Santos, M.V.; Gigante, M.L. Effect of somatic cell count on Prato cheese composition. *J. Dairy Sci.* **2007**, *90*, 630–636. [CrossRef]

31. Franceschi, P.; Malacarne, M.; Formaggioni, P.; Stocco, G.; Cipolat-Gotet, C.; Summer, A. Effect of season and cheese-factory on cheese-making efficiency in Parmigiano Reggiano cheese manufacture. *Foods* **2019**, *8*, 315. [CrossRef] [PubMed]

32. Fleminger, G.; Ragones, H.; Merin, U.; Silanikove, N.; Leitner, G. Chemical and structural characterization of bacterially-derived casein peptides that impair milk clotting. *Int. Dairy J.* **2011**, *21*, 914–920. [CrossRef]

33. Pretto, D.; De Marchi, M.; Penasa, M.; Cassandro, M. Effect of milk composition and coagulation traits on Grana Padano cheese yield under field conditions. *J. Dairy Res.* **2013**, *80*, 1–5. [CrossRef] [PubMed]

34. Verdier-Metz, I.; Coulon, J.B.; Pradel, P. Relationship between milk fat and protein contents and cheese yield. *Anim. Res.* **2001**, *50*, 365–371. [CrossRef]

© 2020 by the authors. Licensee MDPI, Basel, Switzerland. This article is an open access article distributed under the terms and conditions of the Creative Commons Attribution (CC BY) license (http://creativecommons.org/licenses/by/4.0/).

Article

Whey Protein Concentrate as a Novel Source of Bifunctional Peptides with Angiotensin-I Converting Enzyme Inhibitory and Antioxidant Properties: RSM Study

Fatima Abdelhameed Hussein [1,2], Shyan Yea Chay [1], Mohammad Zarei [3], Shehu Muhammad Auwal [4], Azizah Abdul Hamid [1], Wan Zunairah Wan Ibadullah [1] and Nazamid Saari [1,*]

[1]   Department of Food Science, Faculty of Food Science and Technology, University Putra Malaysia, Serdang, Selangor 43400, Malaysia; fatima_abdelhameed@yahoo.com (F.A.H.); shyan_yea@upm.edu.my (S.Y.C.); azizahah@upm.edu.my (A.A.H.); wanzunairah@upm.edu.my (W.Z.W.I.)

[2]   Department of Dairy Production, Faculty of Animal Production, University of Khartoum, PO Box 32, Khartoum North 13314, Sudan

[3]   Department of Food Science and Technology, Faculty of Applied Sciences, Universiti Teknologi MARA, Shah Alam, Selangor 40450, Malaysia; mzarei.mail@gmail.com

[4]   Department of Biochemistry, Faculty of Basic Medical Sciences, Bayero University, Kano 700231, Nigeria; samuhammad.bch@buk.edu.ng

*   Correspondence: nazamid@upm.edu.my; Tel.: +603-9769-8367

Received: 18 September 2019; Accepted: 9 October 2019; Published: 8 January 2020

Abstract: Whey protein concentrate (WPC) is a unique source of protein with numerous nutritional and functional values due to the high content of branched-chain amino acid. This study was designed to establish the optimum conditions for Alcalase-hydrolysis of WPC to produce protein hydrolysates with dual biofunctionalities of angiotensin-I converting enzyme (ACE) inhibitory and antioxidant activities via response surface methodology (RSM). The results showed that the optimum conditions were achieved at temperature = 58.2 °C, E/S ratio = 2.5%, pH = 7.5 and hydrolysis time = 361.8 min in order to obtain the maximum DH (89.2%), ACE-inhibition (98.4%), DPPH● radical scavenging activity (50.1%) and ferrous ion chelation (73.1%). The well-fitted experimental data to predicted data further validates the regression model adequacy. Current study demonstrates the potential of WPC to generate bifunctional hydrolysates with ACE inhibition and antioxidant activity. This finding fosters the use of WPC hydrolysate as a novel, natural ingredient for the development of functional food products.

Keywords: ACE inhibition; antioxidant activity; hydrolysis; response surface methodology; whey protein concentrate

---

1. Introduction

Hypertension is considered a key risk factor in cardiovascular-related diseases including stroke, peripheral and coronary heart diseases. The total number of adults who had elevated blood pressure was 594 million in 1975 and it increased tremendously to 1.13 billion in 2015 [1], attributed to unhealthy diets and poor lifestyle such as alcohol intake and smoking. By 2025, it is projected that hypertension will affect more than 1.5 billion people globally [2]. Angiotensin-I converting enzyme (ACE) has been identified as the key enzyme which increases blood pressure in human body. It cleaves the inactive decapeptide, angiotensin-I, to an octapeptide, angiotensin-II, which acts as a potent vasoconstrictor, as well as inactivates the vasodilating nonapeptide, bradykinin. The dual action of ACE results in a rise in the blood pressure. Inhibition of ACE by reducing angiotensin II generation and increasing bradykinin generation is necessary to control high blood pressure. In recent years, the increasing health awareness drives consumers to look for natural alternatives that are safer with minimal side effects compared to synthetic blood pressure-controlling drugs. Interestingly, positive results have been garnered with the discovery of bioactive peptides derived from various food proteins.

Whey proteins concentrate (WPC), being a natural valuable peptide source, provides considerable nutritional and health benefits for humans due to its high level of branched-chain amino acids, high protein content (30–80%) [3–5] and presence of promising functional molecules including immunoglobulin, α-lactalbumin, lactoperoxidase, albumin, lactoferrin and caseinomacropeptide [6,7]. However, WPC is less studied as compared to whey protein (the crude, non-purified form) and whey protein isolate (higher protein content than WPC). Hydrolysis of these whey products would generate smaller peptide fragments with different bioactivities, such as antioxidative [8], antihypertensive [9–11], immunomodulatory [12], antithrombotic [13], antimicrobial [14] and opiate properties [15]. Of all these bioactivities, ACE inhibition and antioxidant activity are extensively studied, but mainly on a separate basis. For instance, Guo, Pan and Tanokura [9] and O'keeffe, Conesa and FitzGerald [16] reported ACE inhibition between 40.0–84.4% for different whey products while Ajlouni and Pan [17] as well as Zhang et al. [18] reported antioxidant activities between 0.0–72.1%, respectively.

Hydrolysis parameters rely heavily on the purpose of hydrolysis. Different purposes would yield different sets of optimum conditions and enzyme selection. WPC has been widely used to evaluate the optimum hydrolysis conditions using different enzymes. For instance, Guo, Pan and Tanokura [9] reported the optimum temperature = 38.9 °C, pH = 9.2 and E/S ratio = 0.60 for *L.* helveticus LB13 crude proteinase hydrolysis while Naik et al. [19] reported the optimum temperature = 35.5 °C, time = 8 h and pH = 7.3 when Flavourzyme was used. Also, WPC has been hydrolysed using Alcalase, the same enzyme in current work, in several other studies. According to Athira et al. [20], the optimum conditions for Alcalase-assisted WPC hydrolysis were 55 °C, 8 h and pH 9 while Fenoglio et al. [21] suggested to use 50 °C, pH 8–9 and 25 min to achieve optimum hydrolysis. While these studies used the same substrate and same enzyme as in current work, they produced different sets of optimum temperatures, pH and hydrolysis time, depending on the different objectives (i.e., the measured responses) targeted by different works. For instance, Athira et al. [20] focused on single bioactivity of WPC, Fenoglio et al. [21] measured the production of bitter peptides for dessert making while current work measured multiple bioactivities from WPC.

Unlike most works that focused on single bioactivity of WPC, current study evaluates the dual functionality of WPC hydrolysate in terms of ACE inhibition and antioxidant activities by employing 4-factors design using response surface methodology (RSM). RSM is a collection of statistical methods for building a model by setting a combination of factor levels to get an optimum response, by employing a polynomial equation to evaluate the relationship between factors as well as their key and combined effect to induce a desired response [22]. This study serves to determine the optimized conditions to generate WPC hydrolysates that possess maximum ACE-inhibition and antioxidant activities by employing response surface methodology according to Box-Behnken design.

2. Materials and Methods

2.1. Materials

Whey protein concentrate (WPC) was purchased from Purefit Company (Selangor, Malaysia). ACE from rabbit lung, N-Hippuryl-His-Leu hydrate powder and O-phthaldialdehyde (OPA) were obtained from Sigma Chemical Co., (St Louis, MO, USA). Alcalase was purchased from Novozymes (Copenhagen, Denmark). All other chemicals employed were of analytical grade and obtained from either Acros Organics (Geel, Belgium), Merck KGaA (Darmstadt, Germany), Fisher Scientific (Waltham, MA, USA) or J.T. Baker (Phillipsburg, NJ, USA).

2.2. Proximate Analysis of WPC

The proximate analysis was determined according to the method of Association of Official Analytical Chemists [23]. Oven method was employed to measure the moisture content using a forced draft oven (model: Quimis Q-314M242, serial 020, São Paulo, Brazil) at 105 °C until reaching a constant weight. Ash was determined via incineration with a muffle furnace at 550 °C. Soxhlet extraction method with ethylic ether was employed to determine the fat content. The micro Kjeldahl method was used to determine the protein content, whereby a conversion of 6.38 was multiplied. All analyses were performed in triplicates.

2.3. Sodium Dodecyl Sulphate Polyacrylamide Gel Electrophoresis (SDS-PAGE)

SDS-PAGE was performed on WPC hydrolysate as described by Laemmli [24], using 40% acrylamide for protein separation. The hydrolysates were adjusted to a 10 mg protein/mL concentration. The sample solutions and the sample buffer (0.5 M Tris-HCl, pH 6.8, containing 4% SDS, 20% glycerol) were mixed at a 3:1 (*v/v*) ratio and heated for 10 min at 95 °C, prior to loading. The concentrations of stacking and resolving gels were 4% and 12%, respectively. The gel was run at 150 V. Aliquots of 10 μL were loaded into individual wells and a constant current was passed through the gel for 1 h to complete the peptide separation. The gel was stained using coomassie blue (5.0 g coomassie brilliant blue R-250 mixed with 500 mL ethanol, 100 mL acetic acid and 400 mL deionized water), distained with distaining solution containing 200 mL of 40% ethanol and 80 mL of 6% acetic acid with 720 mL deionized water, then preserved in 5% acetic acid. The molecular weights were determined by approximation using low range molecular weight standards (Bio-Rad, Hercules, CA, USA).

2.4. Response Surface Methodology (RSM) Design for WPC Hydrolysis

While optimum hydrolysis conditions are recommended by the enzyme manufacturer, these parameters are given in ranges without pinpointing exact values. Optimum conditions are subjective and multifactor-dependent (purpose of study, sample nature and equipment availability). Thus, optimum conditions should be specifically considered on a case-by-case basis, justifying the need to perform optimization study on WPC hydrolysis. Box-Behnken design (BBD) was employed for optimisation of WPC hydrolysis parameters, whereby each parameter had three levels coded as -1, 0, $+1$, based on four independent variables: temperature (°C, x_1), enzyme to substrate ratio (E/S, x_2), pH (x_3) and time (min, x_4). BBD was chosen over central composite design because it is more economic and suitable for current study which manipulates four independent variables and four dependent variables (responses). A total of twenty-seven runs with three replicates of the central point was performed. For each treatment, the mean pertaining to triplicate measurements was considered as responses for degree of hydrolysis (%, y_1), ACE-inhibition (%, y_2), DPPH• (%, y_3) and ferrous ion chelating (%, y_4). Randomisation of experimental runs was performed to mitigate the impacts of unexpected variability to actual responses.

Briefly, Alcalase was mixed with WPC in 100 mL borate buffer at different ranges of temperatures (40–70 °C), enzyme/substrate (E/S) ratios (0.5%–2.5%), pHs (pH 6.5–10) and times (30–480 min), then incubated in a water bath shaker at an agitation rate of 150 rpm. At the end of hydrolysis, the mixture was placed in boiling water (100 °C) for 10 min to inactivate Alcalase. Upon cooling, centrifugation at 10,000 g was performed for 15 min. The supernatant was stored at -20 °C prior to analysis of degree of hydrolysis (DH), biological activities including ACE-inhibition, DPPH•, and ferrous ions chelating activity.

For data analysis, analysis of variance (ANOVA) and regression equation were employed to determine the regression coefficients, statistical significance of the model terms and to fit the predicted mathematical models with that of the experimental data. The least-squares technique was employed to determine the multiple regression coefficients for predicting quadratic and linear polynomial models pertaining to the calculated response variables [25,26]. The general polynomial model for determining the response is given as follows (Equation (1)):

$$y = \beta_0 + \beta_1 x_1 + \beta_2 x_2 + \beta_3 x_3 + \beta_4 x_4 + \beta_{11} x_1^2 + \beta_{22} x_2^2 + \beta_{33} x_3^2 + \beta_{44} x_4^2 + \beta_{12} x_1 x_2 + \\ \beta_{13} x_1 x_3 + \beta_{14} x_1 x_4 + \beta_{23} x_2 x_3 + \beta_{24} x_2 x_4 + \beta_{34} x_3 x_4 \tag{1}$$

whereby, y represents the response; β_0 signifies the offset term for the design; β_1, β_2, β_3 and β_4 denote the regression coefficients defining the linear effect terms; β_{11}, β_{22}, β_{33} and β_{44} signify the quadratic effects; β_{12}, β_{13}, β_{14}, β_{23}, β_{24} and β_{34} represent the interaction effects; and x_1, x_2, x_3, and x_4 represent independent factors in this model.

All linear effects, whether significant or not, are included in the original regression equations (non-reduced model). On the other hand, for square and interaction effects, only significant terms were taken into consideration for reduced polynomial regression pertaining to degree of hydrolysis (y_1, Equation (2)), ACE-inhibition (y_2, Equation (3)), DPPH• radical scavenging (y_3, Equation (4)) and ferrous ion chelating activity (y_4, Equation (5)).

$$y_1 = -116.761 + 3.421x_1 + 55.453x_2 - 6.253x_3 + 0.219x_4 - 0.056x_1 x_1 + 0.36x_1 x_3 - 5.603x_2 x_3 \tag{2}$$

$$y_2 = -57.8367 - 0.943x_1 + 33.1444x_2 + 34.294x_3 + 0.0542x_4 - 0.0286x_1^2 - 3.347x_3^2 \\ - 0.0002x_4^2 + 0.4761x_1 x_3 + 0.0014x_1 x_4 - 4.0186x_2 x_3 \tag{3}$$

$$y_3 = -190.337 + 6.167x_1 - 0.313x_2 + 5.799x_3 + 0.158x_4 - 0.050x_1 x_1 - 0.018x_3 x_4 \tag{4}$$

$$y_4 = -31.8992 + 1.9866x_1 - 5.3641x_2 + 13.9083x_3 + 0.0609x_4 - 0.0174x_1 x_1 - 0.9596x_3 x_3 \\ - 0.0002x_4 x_4 + 0.0157x_2 x_4 \tag{5}$$

For each response (y-value), four important elements, i.e., R^2, R^2-adj, Fisher values (*f*-values) and *p*-values for regression were taken into account to determine the adequacy of the models in predicting experimental values. Based on reduced models, all significant ($p < 0.05$) interaction effects from two independent factors were visualized as three dimensional (3D) surface plots, with the responses (DH, ACE-inhibitory activity, DPPH• radical scavenging and ferrous ion chelating) plotted on y-axis. Drawing of the 3D plots was obtained by maintaining two constant variables at the centre point while modifying the other two variables within the experimental range. The Minitab statistical package (Minitab Inc., State College, PA, USA) was employed for data analysis, optimisation procedure and experimental design conditions. Specifically, for mathematical optimisation, response optimiser was applied to calculate specific optimum levels of concurrent and individual multiple response optimisations, resulting in appropriate response objectives. The experimental data was compared with the predicted values obtained from the equations to test the adequacy of regression equations. Significant differences were identified at $p < 0.05$.

2.5. Degree of Hydrolysis (DH)

OPA method was employed following the procedure of Church et al. [27] and Salami et al. [28] with minor modifications to measure DH at various hydrolysis time. To prepare OPA reagent, three solutions, A, B and C, were prepared separately. Solution A comprised of 7.62 g of sodium tetrahydroborate and 200 mg of sodium dodecyl sulphate (SDS) dissolved in 150 mL deionised water; Solution B consisted of 160 mg of OPA dissolved in 4 mL of ethanol 96%; and Solution C consisted of 400 µL of β-mercaptoethanol, in 50 mL deionised water. Mixing the 3 solutions would yield fresh OPA reagent. Then, 36 µL of sample (protein concentration = 20 mg/mL) was mixed with 270 µL of OPA reagent in a 96-well plate, incubated for 2 min at room temperature, and the absorbance was measured at 340 nm by employing a microplate reader system (model: PowerWave X340, Biotek Instruments Inc., Winooski, VT, USA). The following equation was employed to calculate the degree of hydrolysis according to Adler-Nissen [29]:

$$DH\ (\%) = \{(A_{sample} - A_{protein})/A_{total}\} \times 100 \tag{6}$$

in which, A_{sample} represents the sample absorbance after hydrolysis, $A_{protein}$ signifies the sample absorbance before hydrolysis (negative control) and A_{total} denotes the absorbance of total protein after 24 h complete hydrolysis using 6M HCl.

2.6. Angiotensin Converting Enzyme (Ace) Inhibitory Activity Assay

ACE–inhibitory was determined according to Cushman and Cheung [30] and Ferreira et al. [31] with certain modifications. The assay was performed by first mixing 10 µL of sample (protein concentration = 20 mg/mL) and 10 µL of ACE (100 mU/mL) then incubated for 10 min, followed by addition of 50 µL 0.1 M sodium borate buffer (pH 8.3) comprising of 5 mM HHL and 0.3 M NaCl into the incubated sample. For control and blank, 10 µL and 20 µL of distilled water were respectively added. After incubation for 60 min at 37 °C, the reaction was terminated by introducing 75 µL of 1 M HCl. After the reaction was stopped, addition of 150 µL of pyridine followed by 75 uL of benzene sulphonyl chloride (BSC) was done. Then, the solution was vortexed for 1 min and cooled in ice bath. After cooling, 200 µL of the solution was transferred to a 96-well plate, and the absorbance was measured at 410 nm. ACE inhibition was calculated as follow:

$$ACE\ inhibition\ (\%) = \{(C - S)/(C - B)\} \times 100 \tag{7}$$

where, C is the absorbance of control (ACE + substrate), S is the absorbance of sample (peptide + ACE + substrate) and B is the absorbance of blank (only substrate).

2.7. Determination of Antioxidant Activity

2.7.1. 1, 1-diphenyl-2-picrylhydrazyl Free Radical Scavenging Assay

The percentage for 1, 1-diphenyl-2-picrylhydrazyl (DPPH•) free radical scavenging activity was obtained according to the method of Hwang et al. [32] with some modification. Briefly, 50 µL of WPC hydrolysate (WPCH, protein concentration = 20 mg/mL) was added to100 µL of DPPH• (0.15 mM, 100% methanol) and 50 µL deionised water. At room temperature, the mixture was placed in a dark room for 45 min and then the absorbance was measured at 517 nm. The scavenging ability of WPCH was calculated as follow:

$$DPPH•\ radical\ scavenging\ activity\ (\%) = \{(A_{control} - A_{sample})/A_{control}\} \times 100 \tag{8}$$

whereby, $A_{control}$ represents the absorbance of control and A_{sample} denotes the absorbance of samples at 517 nm.

2.7.2. Ferrous Ion Chelating Activity Assay

The ferrous ion chelating activity of WPCH was measured based on a previously defined method put forward by Dinis, Madeira and Almeida [33] with some changes. WPCH (100 µL) of concentrations between 0.005 mg/mL to 10 mg/mL, was mixed with 185 µL of double distilled water and 5 µL of 2 mmol/L iron dichloride solution, followed by addition of 10 µL of 5 mmol/L ferrozine solution and mixed in a vigorous manner. At room temperature, the mixture was incubated for 10 min, prior to absorbance measurement at 562 nm.

The following equation was employed to determine the chelating effect:

$$Metal\ chelating\ activity\ (\%) = \{(A_{control} - A_{sample})/A_{control}\} \times 100 \tag{9}$$

where, $A_{control}$ represents the absorbance of control and A_{sample} denotes the absorbance of samples at 562 nm.

3. Results and Discussion

3.1. Proximate Analysis

Table 1 shows the proximate composition of WPC with the values of 5.64%, 76.13%, 3.91% and 2.96% for moisture, protein, fat and ash, respectively. Except for moisture and carbohydrate, all values remained comparable to that reported by Adamson [3], Suthar, Jana and Balakrishnan [4] and Whetstine, Croissant and Drake [5]. In particular, the protein content of WPC is high. Therefore, WPC is considered a valuable low-cost protein source to generate hydrolysate (a mixture of bioactive peptides) with dual functionalities upon optimized hydrolysis.

Table 1. Comparison of proximate composition of whey protein concentrate with literature. All values are reported in dry basis (%).

Parameters	Present Study (Mean ± sd)	Literature Data *
Moisture (%)	5.64 ± 0.03	3.8–4.5
Protein (%)	76.13 ± 0.23	74.8–80.0
Crude fat (%)	3.91 ± 0.75	1.8–7.2
Ash (%)	2.96 ± 0.01	2.7–7.5
Carbohydrates (%)	11.93 ± 0.00	3.5–8.9

* Reported data were compiled from three literatures: Adamson [3] Suthar et al. [4] and Whetstine et al. [5]. sd = standard deviation from 3 readings.

3.2. Molecular Weight of WPC Hydrolysates

The size of the generated peptides is essential for ACE inhibitory, antioxidative and other biological activities. SDS-PAGE was performed to investigate the molecular weight distribution of peptides from WPC hydrolysate at different hydrolysis times. Figure 1 shows that Alcalase efficiently degrades major WPC proteins, i.e., β-lactoglobulin and α-lactalbumin (M_w: 15–20 kDa), into peptides of smaller fragments (below 10 kDa) after 30 min of hydrolysis. Also, serum albumin with heavier molecular weight (around 75 kDa), produces no visible band on the gel after 30 min. These observations are in line with the work performed by Pena-Ramos and Xiong [34], who reported a rapid degradation of WPC proteins into smaller peptides by protease A (an endopeptidase obtained from *Bacillus licheniformis*, now widely known as Alcalase).

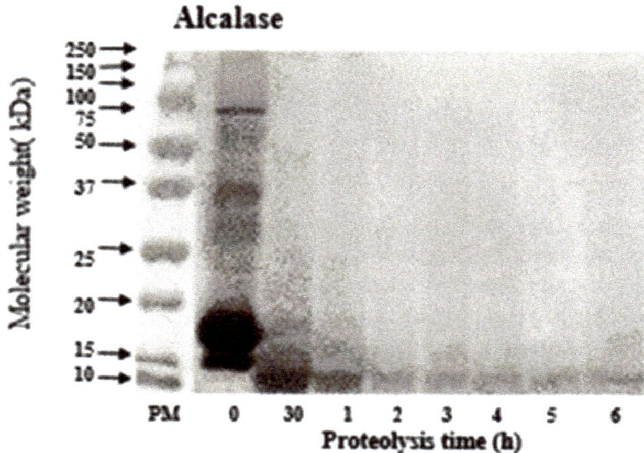

Figure 1. SDS-PAGE of native whey protein concentrate (WPC) and Alcalase-treated WPC at different hydrolysis hours.

3.3. Model Performance Appraisal on the Regression Elements

In terms of statistics, five parameters are crucial, which include estimated regression coefficient (R^2), adjusted coefficient of determination (R^2-adj), lack of fit values, Fisher test value (f-value) and regression p-values. The combination of high R^2, a non-significant lack of fit, high f-value and low p-values signify a good fitness of experimental values to predicted values obtained from regression equations, and that the regression model is statistically sufficient to explain the experimental data. The R^2-adj further improves the reliability of regression model by taking into consideration only the factors (x-values) that significantly affect the response (y-values).

Table 2 presents the predicted and experimental responses of DH (y_1), ACE-inhibitory activity (y_2), DPPH• (y_3) and ferrous ion chelating (y_4) under various hydrolysis conditions, at different combinations of temperature (x_1), E/S ratio (x_2), pH (x_3) and time (x_4). It can be seen that most experimental values were similar to predicted values. From Table 3, the values for R^2 were 91.3% (y_1), 90.0% (y_2), 86.8% (y_3) and 90.8% (y_4) while the result for R^2-adj were 87.4%, 83.7%, 71.4% and 80.1% for y_1, y_2, y_3 and y_4. The lack of fit values for y_1, y_2, y_3 and y_4 were 0.665, 0.060, 0.013 and 0.027 while the f-test values of y_1, y_2, y_3 and y_4 were 23.54, 14.33, 9.80 and 12.33, respectively. These results confirmed the validity of selected regression model to predict the responses (y-values).

3.4. Effects of Temperature, E/S Ratio, pH and Time on Different Responses

3.4.1. Degree of Hydrolysis (DH)

From Table 2, DH between 17.9–93.3% was recorded for hydrolysis under different combinations of temperature, E/S ratio, pH and time. High DH (above 80%) was observed for 7 runs, out of a total of 27 runs, and this is supported by the work of See, Hoo and Babji [35] who reported a DH of 77.03% for fish skin hydrolysis that used a single enzyme of Alcalase. High DH indicates that the selected combinations of temperature, E/S ratio, pH and time were optimum to perform hydrolysis on large, bulky whey protein molecules to effectively cut them into smaller peptides.

The analysis of variance ($p < 0.05$) and regression coefficients for DH are depicted in Table 3. As shown, the regression model for DH, when expressed in full quadratic equation, is significant at $p = 0.000$. This confirms the adequacy of model to fit the DH experimental data. The squares effects were identified as temp*temp and time*time while the interaction effects were found to be temp*pH and E/S*pH. The high R^2, non-significant lack-of-fit and large f-value implies a good fit of experimental values to predicted model. Figure 2 depicts the interaction effects from independent factors, comprised of temperature, E/S ratio and pH, which contributed to the changes in DH. From Figure 2c, DH increased with hydrolysis temperature, reached a maximum point then decreased following higher temperature, forming a quadratic surface plot. This is due to enzyme denaturation of Alcalase at high temperatures which reduces its efficiency to hydrolyse WPC molecules into smaller peptide fragments, and thus results in lower DH values.

Table 2. Behnken design, predicted and experimental values of four response variables (y_1, y_2, y_3, y_4) as affected by temperature (°C), E/S ratio (%), pH and hydrolysis time (min).

Run Order	x_1	x_2	x_3	x_4	% DH (y_1)		%ACE Inhibition (y_2)		% DPPH• (y_3)		% Ferrous Ion Chelating (y_4)	
					Experimental	Predicted	Experimental	Predicted	Experimental	Predicted	Experimental	Predicted
1	40	0.5	8.25	255	30.71	45.22	27.53	26.85	27.53	26.85	73.78	72.09
2	70	0.5	8.25	255	44.63	52.36	51.10	46.28	51.10	46.28	76.30	74.33
3	40	2.5	8.25	255	67.42	63.67	20.97	26.23	20.97	26.23	65.04	69.37
4	70	2.5	8.25	255	62.09	70.81	47.13	45.65	47.13	45.65	74.70	71.61
5	55	1.5	6.5	30	22.67	25.22	31.87	35.59	31.87	35.59	73.48	68.97
6	55	1.5	10	30	42.64	43.29	51.47	54.03	51.47	54.03	58.00	62.23
7	55	1.5	6.5	480	87.35	77.50	49.10	55.03	49.10	55.03	65.63	64.82
8	55	1.5	10	480	91.07	95.57	40.77	45.53	40.77	45.53	58.37	58.08
9	40	1.5	8.25	30	17.90	18.09	34.20	23.81	34.20	23.81	63.44	63.51
10	70	1.5	8.25	30	34.46	25.23	41.33	43.23	41.33	43.23	64.04	65.75
11	40	1.5	8.25	480	72.37	70.37	21.67	29.27	21.67	29.27	60.93	59.36
12	70	1.5	8.25	480	66.52	77.52	51.77	48.7	51.77	48.7	59.00	61.60
13	55	0.5	6.5	255	40.58	42.54	42.80	45.62	42.8	45.62	77.00	77.55
14	55	2.5	6.5	255	79.65	80.60	57.73	45.00	57.73	45.00	74.96	74.83
15	55	0.5	10	255	82.06	80.22	48.33	50.09	48.33	50.09	69.96	70.81
16	55	2.5	10	255	81.91	79.06	42.93	49.46	42.93	49.46	67.59	68.09
17	40	1.5	6.5	255	58.15	54.87	24.5	24.31	24.5	24.31	66.56	71.16
18	70	1.5	6.5	255	51.01	43.09	32.8	43.73	32.8	43.73	73.11	73.40
19	40	1.5	10	255	59.72	54.02	30.4	28.77	30.4	28.77	70.19	64.42
20	70	1.5	10	255	90.43	80.08	51.70	48.20	51.70	48.20	66.22	66.66
21	55	0.5	8.25	30	27.44	25.03	42.37	45.12	42.37	45.12	73.22	73.43
22	55	2.5	8.25	30	35.25	43.48	44.57	44.50	44.57	44.50	65.37	63.64
23	55	0.5	8.25	480	83.48	77.31	60.13	50.59	60.13	50.59	61.22	62.21
24	55	2.5	8.25	480	93.28	95.76	55.17	49.96	55.17	49.96	67.51	66.56
25	55	1.5	8.25	255	60.21	70.61	48.17	47.54	48.17	47.54	75.93	75.76
26	55	1.5	8.25	255	76.95	70.61	49.53	47.54	49.53	47.54	74.93	75.76
27	55	1.5	8.25	255	72.94	70.61	48.30	47.54	48.30	47.54	75.41	75.76

Remark: Independent variables: x_1 = Temperature; x_2 = E/S ratio; x_3 = pH; x_4 = hydrolysis time.

Table 3. Of variance at $p < 0.05$ and regression coefficients for DH (y_1), ACE inhibition (y_2), DPPH• (y_3) and ferrous ion chelating activity (y_4), expressed in full quadratic models.

Factors	y_1	*p*-Value	y_2	*p*-Value	y_3	*p*-Value	y_4	*p*-Value
Regression		0.000		0.000		0.000		0.000
Linear		0.000		0.004		0.000		0.000
Squares		0.001		0.000		0.000		0.000
Interaction		0.013		0.000		0.038		0.024
Constant	−116.761	0.203	−57.8367	0.408	−190.337	0.000	−31.8992	0.355
x_1	3.421	0.110	−0.9430	0.428	6.167	0.000	1.9866	0.003
x_2	55.453	0.010	33.1444	0.007	−0.313	0.865	−5.3641	0.009
x_3	−6.253	0.511	34.2940	0.007	5.799	0.020	13.9083	0.041
x_4	0.219	0.000	0.0542	0.210	0.158	0.027	0.0609	0.001
x_1*x_1	−0.056	0.001	−0.0286	0.003	−0.050	0.000	−0.0174	0.004
x_2*x_2		NS		NS		NS		NS
x_3*x_3		NS	−3.3472	0.000		NS	−0.9596	0.022
x_4*x_4	0.000	0.005	−0.0002	0.000		NS	−0.0002	0.000
x_1*x_2		NS		NS		NS		NS
x_1*x_3	0.360	0.032	0.4761	0.000		NS		NS
x_1*x_4		NS	0.0014	0.050		NS		NS
x_2*x_3	−5.603	0.027	−4.0816	0.006		NS		NS
x_2*x_4		NS		NS		NS	0.0157	0.024
x_3*x_4		NS		NS	−0.018	0.038		NS
R^2	91.27%		89.96%		74.62%		84.56%	
R^2-adj	87.40%		83.68%		67.01%		77.70%	
Lack of fit		0.665		0.060		0.013		0.027
f-value	23.54		14.33		9.80		12.33	

NS means not significant ($p > 0.05$) and thus the respective regression coefficient is not reported.

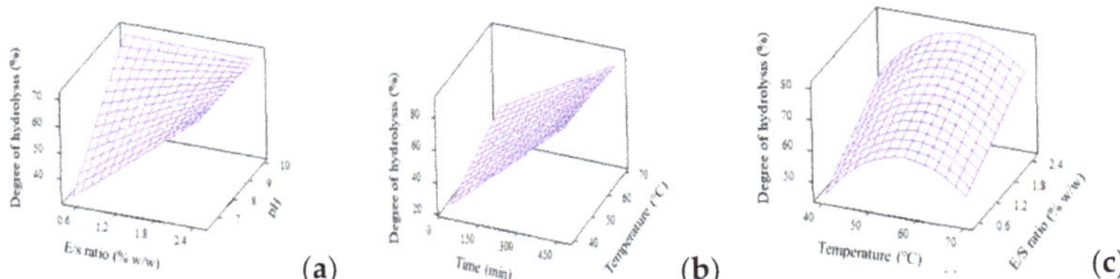

Figure 2. 3D surface plots for degree of hydrolysis (%) as affected by (**a**) E/S ratio and pH; (**b**) Time and temperature and (**c**) Temperature and E/S ratio.

3.4.2. ACE-inhibitory Activity

The analysis of variance ($p < 0.05$) and regression coefficients for ACE-inhibitory activity were generated from reduced full quadratic model and were compiled in Table 3. Six terms were found to be significant to explain the changes in ACE-inhibitory activity, namely, temp*temp, pH*pH, time*time, temp*pH, temp*time and E/S*pH. The regression model for ACE-inhibitory activity with high R^2 of 89.96%, nonsignificant lack-of-fit of 0.06 and large f-value of 14.33 denotes the goodness of fit of experimental results to the predicted ones. The surface plots for ACE-inhibitory activity were constructed based on the interaction effects of temperature, E/S ratio, pH and time (Figure 3).

ACE-inhibitory activity of WPCHs was highly affected by hydrolysis parameters as well as the resulting DH. At higher DH, a higher ACE inhibitory bioactivity was achieved. This result is in line with the work reported by Silvestre et al. [36] and Ghanbari et al. [37], in which the ACE-inhibitory activity of protein hydrolysates was affected by the type of enzymes as well as the degree of hydrolysis. Also, the increased level of E/S ratio and alkali pH resulted in high values of DH to indicate the release of potent peptides which further contributed towards the strong ACE-inhibitory activity. These findings are comparable to another work reported by van der Ven et al. [38] which used the same sample of whey protein hydrolysate.

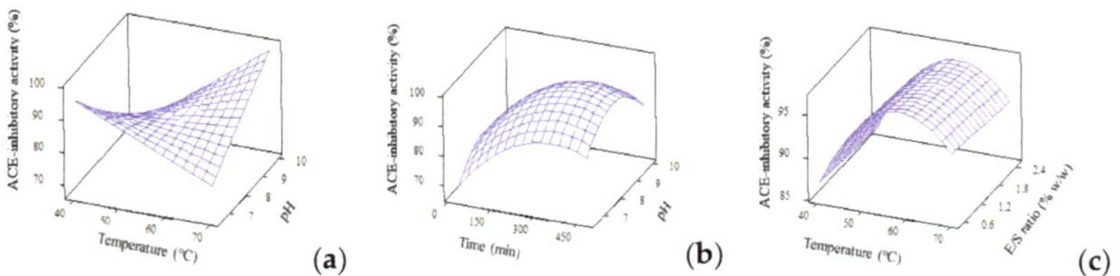

Figure 3. 3D surface plots for ACE-inhibitory activity (%) as affected by (**a**) Temperature and pH; (**b**) Time and pH and (**c**) Temperature and E/S ratio.

3.4.3. Antioxidant Activities

Unlike typical works which perform only single antioxidant assay in RSM study, current work reported two antioxidant activities, i.e., DPPH• radical scavenging and ferrous chelating activities. The reduced quadratic model of DPPH• radical scavenging is shown in Table 3. Even though this model contains only two significant terms, contributed by temp*temp and pH*time, the overall regression equation is significant ($p = 0.000$) due to the strong effect from linear, squares and interaction. Figure 4 presents the 3D plots regarding DPPH• radical scavenging activity. Similar to the effect of temperature on DH, as previously discussed in Section 3.4, the effect of temperature on DPPH• radical scavenging activity is also prominent. As the hydrolysis temperature increased, the scavenging activity increased until a maximum point, then decreased with higher temperature. This is, again, due to the application of temperature which allow/deter optimum Alcalase activity to generate peptides with DPPH• radical scavenging activity. The ability for scavenging free radicals could be attributed to the peptides' ability to donate hydrogen, neutralise or stabilise free radicals and then cease their propagation. This potency is strongly affected by the amino acid composition in samples, especially aromatic amino acids like tryptophan, phenylalanine, histidine and tyrosine as well as hydrophobic amino acids like alanine, leucine and valine, with methionine playing the most critical role on affecting the scavenging activity [39–41].

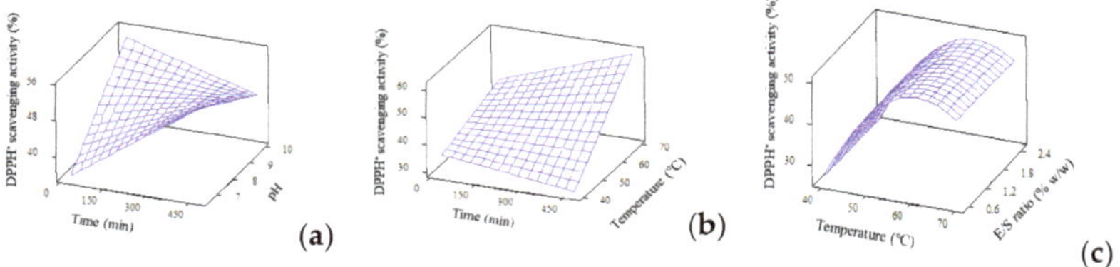

Figure 4. 3D surface plots for DPPH• radical scavenging (%) as affected by (**a**) Time and pH; (**b**) Time and temperature and (**c**) Temperature and E/S ratio.

Ferrous chelating activity, which helps to minimise lipid oxidation, was exhibited by WPC hydrolysate. The quadratic model for this bioactivity is depicted in Table 3. The model is significant ($p = 0.000$) and encompasses all linear, squares, interaction effects with p values of 0.000, 0.000 and 0.024, respectively. Four significant terms were reported for ferrous chelating activity, which consisted of mainly squares effect, i.e., temp*temp, pH*pH, time*time, along with an interaction effect of E/S*time. The 3D plots for ferrous chelating activity are shown in Figure 5. The potency of peptides are thought to be affected by sources of protein, type of proteases, amino acid composition and time of hydrolysis [42,43]. The higher ferrous ion chelating activity demonstrated by WPCH proves the suitability of Alcalase as the choice of enzyme to produce biopeptide molecules.

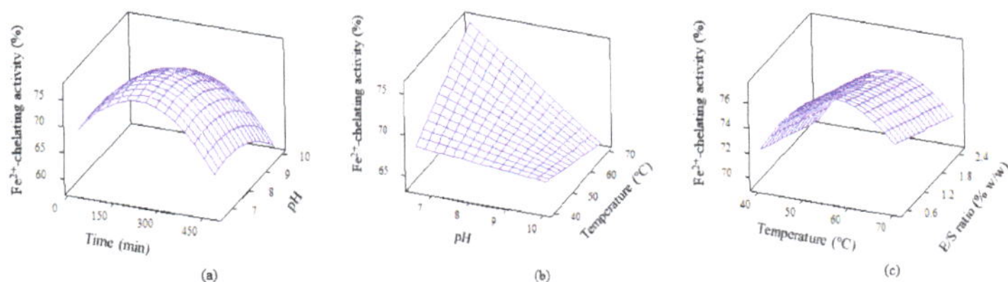

Figure 5. 3D surface plots for ferrous chelating activity as affected by (**a**) Time and pH; (**b**) pH and temperature and (**c**) Temperature and E/S ratio.

3.5. Optimized Hydrolysis Condition For WPC and Model Validation

Optimization procedure was performed using Box-Behnken design. Figure 6 presents the optimum predicted hydrolysis conditions for WPC that could theoretically produces the maximum DH, ACE-inhibitory, DPPH• radical scavenging and ferrous ion chelating activities. The predicted optimal conditions were as follow: temperature = 58.2 °C, E/S ratio = 2.5%, pH = 7.5, and hydrolysis time = 361.8 min. The predicted parameters were employed in actual experiments for model validation purpose. The model was confirmed to be validate via two observations: (i) The overall desirability (D-value) of 0.88 indicated a high level of confidence for the model to produce the responses as predicted; and (ii) The experimental values for DH (89.2% vs. theoretical 89.6%), ACE-inhibitory activity (98.4% vs. theoretical 98.8%), DPPH• radical scavenging (50.1% vs. theoretical 50.6%) and ferrous ion chelating (73.1% vs. theoretical 74.0%) were identical to their respective predicted values. On the other hand, the goodness-of-fit between experimental and predicted data is depicted in Figure 7a–d. In all figures, the data points fall closely to the regression line. Also, high R^2 of 91.3%, 90.0%, 74.6% and 84.6% for DH, ACE inhibitory, DPPH• scavenging and ferrous ion chelating, respectively, indicated that most of the variabilities of these responses were explained by their linear models.

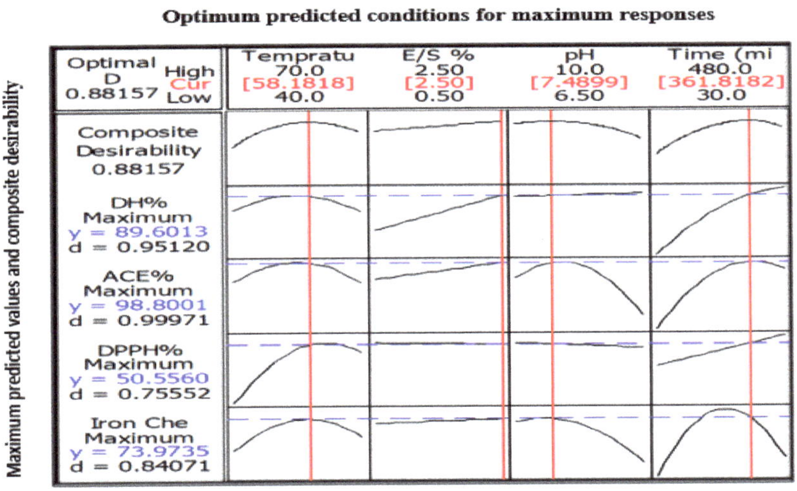

Optimum predicted conditions for maximum responses

		Tempratu	E/S %	pH	Time (mi
Optimal	High	70.0	2.50	10.0	480.0
D	Cur	[58.1818]	[2.50]	[7.4899]	[361.8182]
0.88157	Low	40.0	0.50	6.50	30.0
Composite Desirability 0.88157					
DH% Maximum y = 89.6013 d = 0.95120					
ACE% Maximum y = 98.8001 d = 0.99971					
DPPH% Maximum y = 50.5560 d = 0.75552					
Iron Che Maximum y = 73.9735 d = 0.84071					

Maximum predicted values and composite desirability

Figure 6. The predicted maximum responses (DH, ACE-inhibitory, DPPH• radical scavenging, ferrous ion chelating activity) based on predicted optimum hydrolysis parameters (temperature, E/S ratio, pH and hydrolysis time). Composite desirability = overall desirability encompassing all responses and d = individual desirability. Optimum hydrolysis parameters are denoted as red lines while the maximum responses are denoted as blue dotted lines.

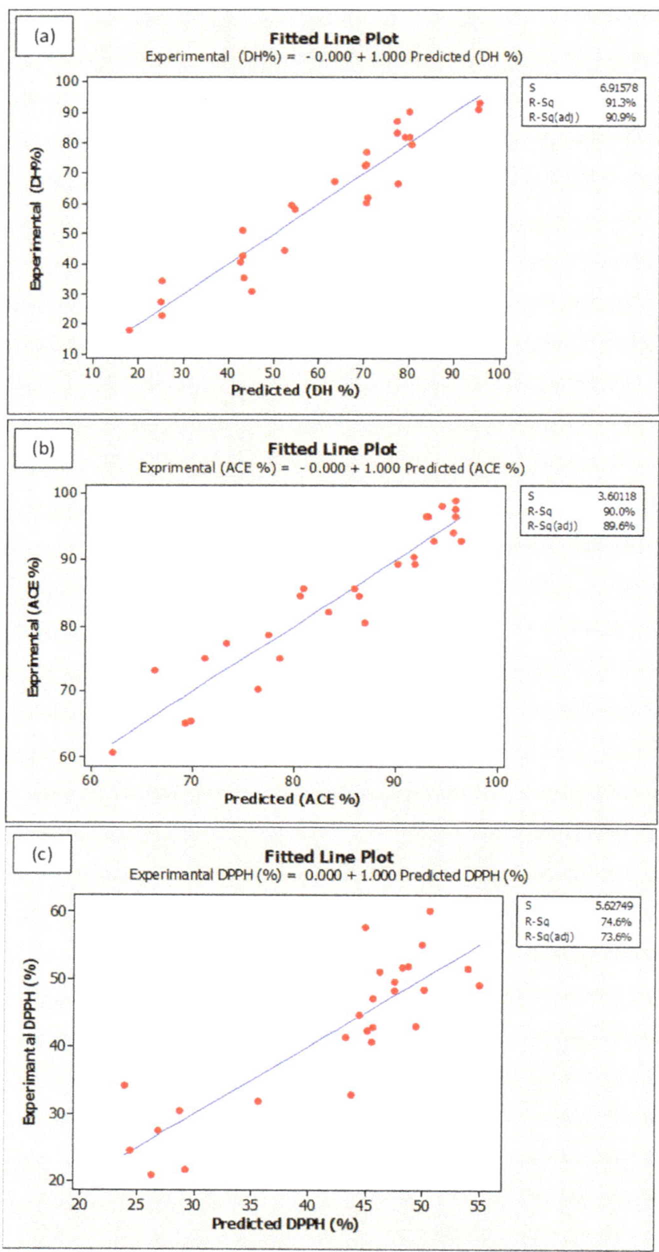

Figure 7. *Cont.*

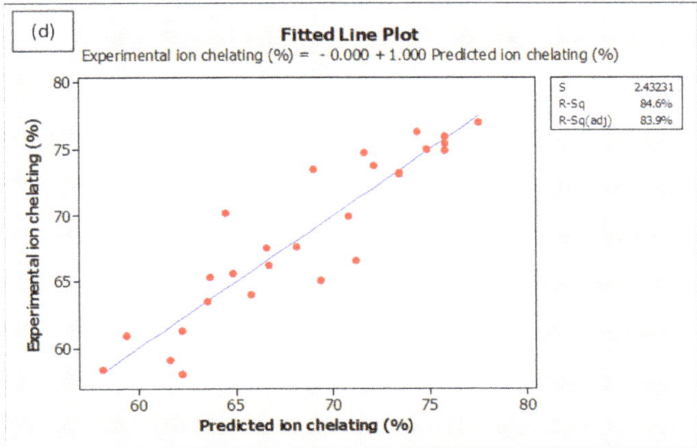

Figure 7. Fitted line plots for predicted and experimental values for (**a**) degree of hydrolysis, (**b**) ACE-inhibition, (**c**) DPPH• radical scavenging activity and (**d**) ferrous ion chelating activity. All values are reported in %.

4. Conclusions

This study represents a pioneer work on the optimization of whey protein concentrate hydrolysis using Alcalase to generate biopeptides with dual functionalities (ACE inhibition and antioxidant activities). The optimized hydrolysis parameters were determined as follow: Temperature = 58.2 °C, E/S ratio = 2.5%, pH = 7.5 and hydrolysis time = 361.8 min. Under such conditions, the experimental values did not show any statistical differences from the predicted values but fitted closely to them. Results showed that the selected hydrolysis parameters were able to produce whey protein hydrolysate with dual bioactivities of ACE inhibition and antioxidant at desirable levels. This study proved the reliability of the selected regression models to sufficiently explain the factor-response relationship during WPC hydrolysis and that the predicted optimum hydrolysis conditions are valid to generate, from WPC, bioactive peptides with ACE inhibitory and antioxidative activities. Further work on the in vivo efficacy and subsequent clinical trials of WPC hydrolysate is deemed appropriate.

Author Contributions: Conceptualization: A.A.H., N.S.; data curation: S.Y.C., M.Z.; investigation: F.A.H.; methodology: S.M.A., W.Z.W.I.; supervision and resources: N.S., writing—original draft: F.A.H., S.Y.C., M.Z., S.M.A. writing—review and editing: A.A.H., W.Z.W.I., N.S. All authors have read and agreed to the published version of the manuscript.

Funding: This research is funded by Universiti Putra Malaysia under project GP-IPS/2018/9643400 (vot no: 9643400) and the article processing charges is funded by Universiti Putra Malaysia.

Conflicts of Interest: All authors declare no conflict of interest.

References

1. Zhou, B.; Bentham, J.; Di Cesare, M.; Bixby, H.; Danaei, G.; Cowan, M.J.; Paciorek, C.J.; Singh, G.; Hajifathalian, K.; Bennett, J.E. Worldwide trends in blood pressure from 1975 to 2015: A pooled analysis of 1479 population-based measurement studies with 19 1 million participants. *Lancet* **2017**, *389*, 37–55. [CrossRef]

2. Kearney, P.M.; Whelton, M.; Reynolds, K.; Muntner, P.; Whelton, P.K.; He, J. Global burden of hypertension: Analysis of worldwide data. *Lancet* **2005**, *365*, 217–223. [CrossRef]

3. Adamson, N. Whey processing. In *Dairy Processing Handbook*; Bylund, G., Ed.; Tetra Pak Processing Systems: Lund, Sweden, 2015; pp. 331–352.

4. Suthar, J.; Jana, A.; Balakrishnan, S. High protein milk ingredients-A tool for value-addition to dairy and food products. *J. Dairy Vet. Anim. Res.* **2017**, *6*, 00171.

5.   Whetstine, M.C.; Croissant, A.; Drake, M. Characterization of dried whey protein concentrate and isolate flavor. *J. Dairy Sci.* **2005**, *88*, 3826–3839. [CrossRef]

6.   del Mar Contreras, M.; Hernández-Ledesma, B.; Amigo, L.; Martín-Álvarez, P.J.; Recio, I. Production of antioxidant hydrolyzates from a whey protein concentrate with thermolysin: Optimization by response surface methodology. *LWT Food Sci. Technol.* **2011**, *44*, 9–15. [CrossRef]

7.   Muro Urista, C.; Álvarez Fernández, R.; Riera Rodriguez, F.; Arana Cuenca, A.; Tellez Jurado, A. Production and functionality of active peptides from milk. *Food Sci. Technol. Int.* **2011**, *17*, 293–317. [CrossRef] [PubMed]

8.   Lin, S.; Tian, W.; Li, H.; Cao, J.; Jiang, W. Improving antioxidant activities of whey protein hydrolysates obtained by thermal preheat treatment of pepsin, trypsin, alcalase and flavourzyme. *Int. J. Food Sci. Technol.* **2012**, *47*, 2045–2051. [CrossRef]

9.   Guo, Y.; Pan, D.; Tanokura, M. Optimisation of hydrolysis conditions for the production of the angiotensin-I converting enzyme (ACE) inhibitory peptides from whey protein using response surface methodology. *Food Chem.* **2009**, *114*, 328–333. [CrossRef]

10.   Morais, H.A.; Silvestre, M.P.C.; Silva, M.R.; Silva, V.D.M.; Batista, M.A.; e Silva, A.C.S.; Silveira, J.N. Enzymatic hydrolysis of whey protein concentrate: Effect of enzyme type and enzyme: Substrate ratio on peptide profile. *J. Food Sci. Technol.* **2015**, *52*, 201–210. [CrossRef]

11.   Tavares, T.; Malcata, F. Whey proteins as source of bioactive peptides against hypertension. In *Bioactive Food Peptides in Health and Disease*; Hernández-Ledesma, B., Ed.; InTech Publisher: London, UK, 2013; pp. 75–114.

12.   Wu, Q.; Zhang, X.; Jia, J.; Kuang, C.; Yang, H. Effect of ultrasonic pretreatment on whey protein hydrolysis by alcalase: Thermodynamic parameters, physicochemical properties and bioactivities. *Process Biochem.* **2018**, *67*, 46–54. [CrossRef]

13.   Silveira, S.T.; Martínez Maqueda, D.; Recio, I.; Hernández Ledesma, B. Dipeptidyl peptidase-IV inhibitory peptides generated by tryptic hydrolysis of a whey protein concentrate rich in β-lactoglobulin. *Food Chem.* **2013**, *141*, 1072–1077. [CrossRef] [PubMed]

14.   Boyacı, D.; Korel, F.; Yemenicioğlu, A. Development of activate-at-home-type edible antimicrobial films: An example pH-triggering mechanism formed for smoked salmon slices using lysozyme in whey protein films. *Food Hydrocoll.* **2016**, *60*, 170–178. [CrossRef]

15.   Madureira, A.; Tavares, T.; Gomes, A.M.P.; Pintado, M.; Malcata, F.X. Invited review: Physiological properties of bioactive peptides obtained from whey proteins. *J. Dairy Sci.* **2010**, *93*, 437–455. [CrossRef] [PubMed]

16.   O'keeffe, M.B.; Conesa, C.; FitzGerald, R.J. Identification of angiotensin converting enzyme inhibitory and antioxidant peptides in a whey protein concentrate hydrolysate produced at semi-pilot scale. *Int. J. Food Sci. Technol.* **2017**, *52*, 1751–1759. [CrossRef]

17.   Ajlouni, S.; Pan, Y. Effect of pH and whey protein isolate to glucose ratios on the formation of Maillard reaction products as antioxidants. *Kasetsart J. (Nat. Sci.)* **2014**, *48*, 759–768.

18.   Zhang, Q.X.; Wu, H.; Ling, Y.F.; Lu, R.R. Isolation and identification of antioxidant peptides derived from whey protein enzymatic hydrolysate by consecutive chromatography and Q-TOF MS. *J. Dairy Res.* **2013**, *80*, 367–373. [CrossRef]

19.   Naik, L.; Mann, B.; Bajaj, R.; Sangwan, R.; Sharma, R. Process optimization for the production of bio-functional whey protein hydrolysates: Adopting response surface methodology. *Int. J. Pept. Res.* **2013**, *19*, 231–237. [CrossRef]

20.   Athira, S.; Mann, B.; Saini, P.; Sharma, R.; Kumar, R.; Singh, A.K. Production and characterisation of whey protein hydrolysate having antioxidant activity from cheese whey. *J. Sci. Food Agric.* **2015**, *95*, 2908–2915. [CrossRef]

21.   Fenoglio, C.; Vierling, N.; Manzo, R.; Ceruti, R.; Sihufe, G.; Mammarella, E. Whey protein hydrolysis with free and immobilized alcalase®: Effects of operating parameters on the modulation of peptide profiles obtained. *Am. J. Food Technol.* **2016**, *11*, 152–158.

22.   Bezerra, M.; Santelli, R.; Oliveira, E.; Villar, L.; Escaleira, L. Response surface methodology (RSM) as a tool for optimization in analytical chemistry. *Talanta* **2008**, *76*, 965–977. [CrossRef]

23.   AOAC International. *Official Methods of Analysis of AOAC International*, 18th ed.; Association of Official Analytical Chemists (AOAC) International: Rockville, MD, USA, 2005.

24.   Laemmli, U.K. Cleavage of structural proteins during the assembly of the head of bacteriophage T4. *Nature* **1970**, *227*, 680. [CrossRef] [PubMed]

25.   Alexopoulos, E. Introduction to multivariate regression analysis. *Hippokratia* **2010**, *14*, 23–28. [PubMed]

26. Mayer, R.; Montgomery, D. *Response Surface Methodology: Process and Product Optimization Using Designed Experiments*; John Wiley and Sons, Inc.: New York, NY, USA, 1995.
27. Church, F.; Swaisgood, H.; Porter, D.; Catignani, G. Spectrophotometric assay using o-phthaldialdehyde for determination of proteolysis in milk and isolated milk proteins. *J. Dairy Sci.* **1983**, *66*, 1219–1227. [CrossRef]
28. Salami, M.; Yousefi, R.; Ehsani, M.R.; Dalgalarrondo, M.; Chobert, J.M.; Haertlé, T.; Razavi, S.H.; Saboury, A.A.; Niasari-Naslaji, A.; Moosavi-Movahedi, A.A. Kinetic characterization of hydrolysis of camel and bovine milk proteins by pancreatic enzymes. *Int. Dairy J.* **2008**, *18*, 1097–1102. [CrossRef]
29. Adler-Nissen, J. *Enzymic Hydrolysis of Food Proteins*; Elsevier Applied Science Publishers: New York, NY, USA, 1986.
30. Cushman, D.; Cheung, H. Spectrophotometric assay and properties of the angiotensin-converting enzyme of rabbit lung. *Biochem. Pharm.* **1971**, *20*, 1637–1648. [CrossRef]
31. Ferreira, I.; Pinho, O.; Mota, M.; Tavares, P.; Pereira, A.; Goncalves, M.; Torres, D.; Rocha, C.; Teixeira, J. Preparation of ingredients containing an ACE-inhibitory peptide by tryptic hydrolysis of whey protein concentrates. *Int. Dairy J.* **2007**, *17*, 481–487. [CrossRef]
32. Hwang, J.Y.; Shyu, Y.S.; Wang, Y.T.; Hsu, C.K. Antioxidative properties of protein hydrolysate from defatted peanut kernels treated with esperase. *LWT Food Sci. Technol.* **2010**, *43*, 285–290. [CrossRef]
33. Dinis, T.C.; Madeira, V.M.; Almeida, L.M. Action of phenolic derivatives (acetaminophen, salicylate, and 5-aminosalicylate) as inhibitors of membrane lipid peroxidation and as peroxyl radical scavengers. *Arch. Biochem. Biophys.* **1994**, *315*, 161–169. [CrossRef]
34. Pena-Ramos, E.; Xiong, Y. Antioxidative activity of whey protein hydrolysates in a liposomal system. *J. Dairy Sci.* **2001**, *84*, 2577–2583. [CrossRef]
35. See, S.F.; Hoo, L.L.; Babji, A.S. Optimization of enzymatic hydrolysis of Salmon (*Salmo salar*) skin by Alcalase. *Int. Food Res. J.* **2011**, *18*, 1359–1365.
36. Silvestre, M.P.C.; Silva, M.R.; Silva, V.D.M.; Souza, M.W.S.; Junior, L.; Oliveira, C.; Afonso, W.O. Analysis of whey protein hydrolysates: Peptide profile and ACE inhibitory activity. *Braz. J. Pharm. Sci.* **2012**, *48*, 747–757. [CrossRef]
37. Ghanbari, R.; Zarei, M.; Ebrahimpour, A.; Abdul-Hamid, A.; Ismail, A.; Saari, N. Angiotensin-I converting enzyme (ACE) inhibitory and anti-oxidant activities of sea cucumber (*Actinopyga lecanora*) hydrolysates. *Int. J. Mol. Sci.* **2015**, *16*, 28870–28885. [CrossRef] [PubMed]
38. van der Ven, C.; Gruppen, H.; de Bont, D.; Voragen, A. Optimisation of the angiotensin converting enzyme inhibition by whey protein hydrolysates using response surface methodology. *Int. Dairy J.* **2002**, *12*, 813–820. [CrossRef]
39. Chi, C.F.; Hu, F.Y.; Wang, B.; Li, Z.R.; Luo, H.Y. Influence of amino acid compositions and peptide profiles on antioxidant capacities of two protein hydrolysates from skipjack tuna (*Katsuwonus pelamis*) dark muscle. *Mar. Drugs* **2015**, *13*, 2580–2601. [CrossRef]
40. Luo, S.; Levine, R.L. Methionine in proteins defends against oxidative stress. *FASEB J.* **2009**, *23*, 464–472. [CrossRef]
41. Zou, T.B.; He, T.P.; Li, H.B.; Tang, H.W.; Xia, E.Q. The structure-activity relationship of the antioxidant peptides from natural proteins. *Molecules* **2016**, *21*, 72. [CrossRef]
42. Karamać, M.; Kosińska-Cagnazzo, A.; Kulczyk, A. Use of different proteases to obtain flaxseed protein hydrolysates with antioxidant activity. *Int. J. Mol. Sci.* **2016**, *17*, 1027. [CrossRef]
43. O'Loughlin, I.B.; Kelly, P.M.; Murray, B.A.; FitzGerald, R.J.; Brodkorb, A. Molecular characterization of whey protein hydrolysate fractions with ferrous chelating and enhanced iron solubility capabilities. *J. Agric. Food Chem.* **2015**, *63*, 2708–2714. [CrossRef]

© 2020 by the authors. Licensee MDPI, Basel, Switzerland. This article is an open access article distributed under the terms and conditions of the Creative Commons Attribution (CC BY) license (http://creativecommons.org/licenses/by/4.0/).

MDPI
St. Alban-Anlage 66
4052 Basel
Switzerland
Tel. +41 61 683 77 34
Fax +41 61 302 89 18
www.mdpi.com

Foods Editorial Office
E-mail: foods@mdpi.com
www.mdpi.com/journal/foods

www.ingramcontent.com/pod-product-compliance
Lightning Source LLC
LaVergne TN
LVHW080557180726
843515LV00004BB/514